Eco-Materials Processing and Design

Eco-Materials Processing and Design

Ceramic Transactions Series, Volume 193

Proceedings of the 6th Pacific Rim Conference on Ceramic and Glass Technology (PacRim6); September 11–16, 2005; Maui, Hawaii

Edited by
Koji Watari
Soo-Wohn Lee

The American Ceramic Society

WILEY-INTERSCIENCE

A JOHN WILEY & SONS, INC., PUBLICATION

Published by John Wiley & Sons, Inc., Hoboken, New Jersey
Published simultaneously in Canada.

For general information on our other products and services please contact our Customer Care Department within the U.S. at 877-762-2974, outside the U.S. at 317-572-3993 or fax 317-572-4002.

Wiley also publishes its books in a variety of electronic formats. Some content that appears in print, however, may not be available in electronic format.

Library of Congress Cataloging-in-Publication Data is available.

ISBN-13 978-0-470-08050-4
ISBN-10 0-470-08050-7

10 9 8 7 6 5 4 3 2 1

MIX
Paper from
responsible sources
FSC
www.fsc.org
FSC® C013604

Contents

Preface

The following book contains selected papers presented in two technical sessions entitled "Green Manufacturing in Ceramics—Solutions and Developments for Global Environmental Problems," and "Research and Development of Photocatalysts." These sessions were presented at the 6th Pacific Rim Conference on Ceramic and Glass Technology (PacRim 6), which was held September 11-16, 2005, in Maui, Hawaii.

Because of serious global environmental problems, the ceramic manufacturing industries have been required to pay close attention to the formation and accumulation of large volumes of carbon dioxide and other poisonous gases, the reduction of raw materials, the consumption of energy, and other serious issues. From the background, we direct Eco-Material Processing and Design for advanced ceramics, such as room-temperature and low-temperature synthesis, low-energy processing, aqueous synthesis and processing, reusing and recycling wasted materials, and elimination and reduction of hazardous materials. Furthermore, the ceramic industries are focused on solving significant environmental problems by making the best use of the distinguishing characteristics of all the materials. Titanium dioxide has been found to be a key ceramic for contributing to environmental purification and for creating new clean energy. Such research and development based on Eco-Material Processing and Design are indispensable to ceramic scientists and engineers.

This collection of high-quality research papers presents current developments in the area of Eco-Material Processing and Design. Through the publication of the following book, we hope that those who were unable to attend this interesting and timely symposium will benefit from its contents.

KOJI WATARI
National Institute of Advanced Industrial Science and Technology (AIST)

SOO WOHN LEE
Sun Moon University

Eco-Materials Processing and Design
Edited by Koji Watari and Soo-Wohn Lee

PREPARATION OF SIALON-BASED MATERIALS FROM COAL FLY ASH USING CARBOTHERMAL REDUCTION AND NITRIDATION METHOD

Masahiro Kamiya, Ryo Sasai and Hideaki Itoh
EcoTopia Science Institute, Nagoya University
Furo-cho, Chikusa-ku, Nagoya 464-8603, Japan

ABSTRACT

Sialon-based porous material was prepared from coal fly ash by carbothermal reduction and nitridation method. Coal fly ash discharged from a thermal power plant was used as a starting material, and graphite was used as a carbon reductant. The coal fly ash and graphite with or without Al_2O_3 for composition adjustment were wet-ball-milled, dried and die-pressed into a cylindrical shape. The green compact was fired under N_2 flow in a temperature range of 1200-1450°C for 0-2 h, after which the residual graphite in the sample was removed by oxidation in air at 800°C for 1 h. A bulk porous material resulted from the calcination at 1450°C for 2 h, was mainly composed of β-sialon and a small amount of hematite and iron silicide. X-ray diffraction analysis showed that quartz and mullite phases, originally contained in the coal fly ash, firstly began to decompose into intermediate phases such as χ-sialon, Si_2N_2O and Al_2O_3 during firing, and simultaneously β-sialon phase was formed through the above intermediate phases. The porous materials obtained with and without Al_2O_3 had a porosity of around 55-65%, which depended on the preparation conditions.

INTRODUCTION

Coal is regarded as an oil alternative energy source from the viewpoints of stable energy supply and its low cost, which ranks second to atomic power, and the amount of coal thermal power generation tends to increase every year in Japan. On the other hand, the coal combustion accompanies a large amount of coal ash as a by-product. About ten million tons of the coal ash was discharged as an inorganic solid waste in 2003. The fly ashes are mainly used in cement and concrete manufacturing, because the fly ashes are mainly composed of SiO_2 and Al_2O_3. However, it is insufficient to reduce the amount of waste only for this purpose. Therefore, it is essential to develop a newly recyclable route of the fly ash.

Sialon-based materials have attracted great attention because of their excellent properties, such as high strength, high toughness and chemical inertness at elevated temperatures.[1] Generally, sialons are prepared from sophisticated starting materials such as α- and β-Si_3N_4, AlN, Al_2O_3, SiO_2 and so on.[1] The use of such materials allows preparing high performance sialons, but the price of the product is so high that it limits more versatile industrial application of sialon. In order to reduce the price of the sialon products, investigations have been performed to synthesize the materials on the basis of carbothermal conversion in a flow of N_2 from clay instead of relatively expensive starting materials.[2, 3]

Recently, porous ceramics are applied to hot-gas filters, water cleaners, catalyst supports, and, and so on. A porous structure can be obtained through a conventional powder-processing route with the incorporation of some pore-forming agents such as sawdust, starch, graphite or organic particulates. In this study, sialon-based porous materials were prepared from the coal fly ash (CFA) with graphite particles as the pore-forming agent and reductant using the carbothermal reduction and nitridation method. Crystal phases, microstructure, porosity, pore size distribution for the resultants were investigated.

EXPERIMENTAL

Coal fly ash (CFA) discharged from a thermal power plant in Japan was used as a starting material. Graphite (purity 99.7%, particle size 5μm, Kojundo Chemical Lab. Co., LTD.) was used as a carbon reductant. Coal fly ash and graphite with or without Al_2O_3 (purity 99.99%, particle size 0.2μm, TM-D, TAIMEI Chemicals Co., LTD.) for composition adjustment were planetary ball-milled in ethanol for 3 h, after which the mixed slurry was vacuum-dried and then the dried powder was pressed uniaxially into a cylindrical shape. The green compact was fired under N_2 flow in a temperature range of 1200-1450°C for 0-2 h. Then, the residual graphite in the sample was removed by oxidation in air at 800°C for 1 h.

Crystalline phases in the coal fly ash and the calcined specimens were identified by X-ray Diffraction (XRD, RINT2500H, Rigaku). Minor metallic components in the coal fly ash were quantified using Inductively Coupled Plasma-Atomic Emission Spectrometry (ICP-AES, 3300DV, Perkin Elmer Japan). Microstructure observation was performed by a Scanning Electron Microscope (SEM, JSM-6330F, JEOL Co.) equipped with an Energy Dispersive Spectrometer (EDS, JED-2140, JEOL Co.). Porosity and pore size distribution of the calcined samples were measured by Archimedes method and Mercury Porosimetry (PORE MASTER 33, Quanta Chrome Co.), respectively.

RESULTS AND DISCUSSION

Coal fly ash study

The mean chemical composition of the coal fly ash used in this study is shown in Table 1. The fly ash was mainly composed of silicon and aluminum oxides. Minor oxide components such as sodium, potassium, calcium, manganese, titanium and iron were detected, and the amount of the each component was a few mass%. The loss on ignition was assumed to correspond with the amount of unburned carbon residues in the sample.

Figure 1 shows the SEM micrograph of the coal fly ash. Fly ashes are spherical powders with a diameter below several tens of micrometers. XRD pattern in Fig. 2 showed that the coal fly ash was mainly composed of quartz (SiO_2) and mullite ($Al_6Si_2O_{13}$) phases. In addition, a broad peak around 25° was observed, which would be attributed to the existence of a glass phase such as aluminosilicate glass.

Table 1. Chemical composition of coal fly ash.

	SiO_2	Al_2O_3	Na_2O	K_2O	CaO	MgO	TiO_2	Fe_2O_3	LOI*
mass%	59.2	29.1	2.4	1.7	2.9	1.3	1.5	4.6	2.6
mmol/1g-CFA	0.99	0.29	0.04	0.02	0.05	0.03	0.02	0.03	

*Loss on Ignition.

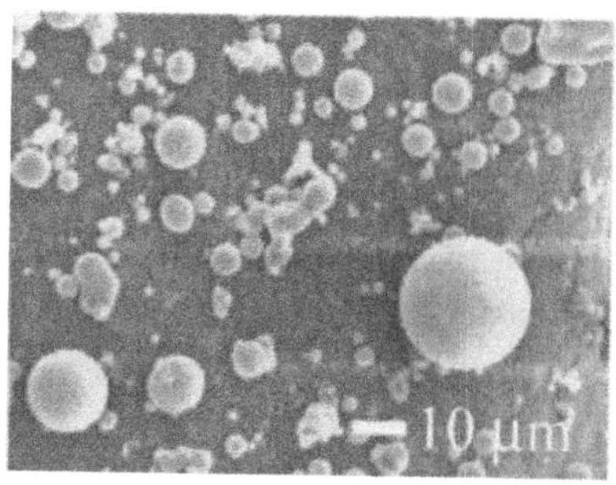

Figure 1. SEM micrograph of the coal fly ash.

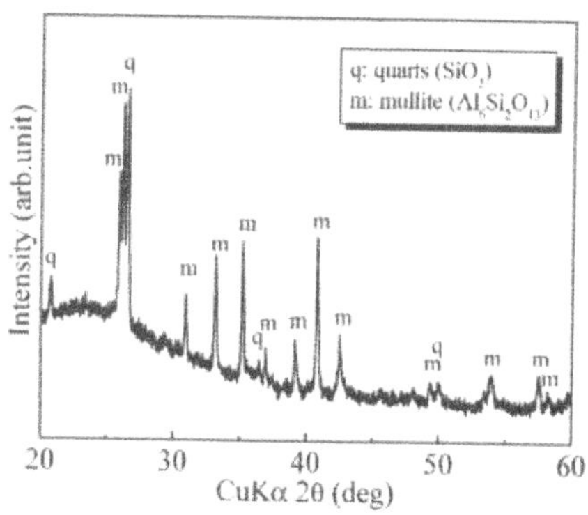

Figure 2. XRD pattern of the coal fly ash.

Table 2 shows the mineralogical composition of the coal fly ash determined by the XRD examination using internal standards. This result indicates that greater part of the silicon and aluminum is included in the glass phase, which corresponds to around 87% of the total silicon contained in the fly ash, and the remaining 13% is in the crystalline phase; 3% is as a form of quartz and 10% is as that of mullite. On the other hand, the half of the total aluminum is included in both the glass phase and the mullite phase, respectively.

Table 2. Mineralogical composition of the coal fly ash.

Crystal phase (mass%)		Glass phase* (mass%)	
Quartz	Mullite	SiO_2	Al_2O_3
1.7	21.7	51.4	13.5

*The amount of glass phase in the coal fly ash was calculated from the chemical composition and the crystal phase amount.

Preparation of sialons

Two kinds of powder compacts with different compositions were prepared; one was the sample A, which had no compositional adjustment by Al_2O_3, whereas another one was the sample B, in which the compensating quantity of Al_2O_3 was added based on the amounts of SiO_2 and Al_2O_3 present in the coal fly ash by referring to the following equation,

$$6SiO_2 + 3Al_2O_3 + 15C + 5N_2 \rightarrow 2Si_3Al_3O_3N_5 + 15CO. \quad (1)$$

The composition of the mixed powders and the porosity of the green bodies are listed in Table 3. As a result of calculation of the porosities, samples A and B were expected to have a porosity of around 50%.

Table 3. Composition of the mixed powders and porosity of the powder compacts.

Sample	Molar ratio,	Mass%			Vol%*			Porosity**,%
	$SiO_2/Al_2O_3/C$	CFA	Al_2O_3	C	CFA	Al_2O_3	C	Green body
A	1.0/0.3/2.5	77	0	23	78	0	22	53
B	1.0/0.5/2.5	66	14	20	71	8	20	52

*Calculated using the density of CFA (2.2g/cm^3), α-Al_2O_3 and graphite.

** $\text{Porosity} = 100 - (V_{CFA} + V_{Al_2O_3})/V_{bulk} \times 100$. V_{CFA}, $V_{Al_2O_3}$ and V_{bulk} are the volume of coal fly ash, α-Al_2O_3 and green body, respectively.

Sialon-based porous materials were prepared from the powder compacts using carbothermal reduction and nitridation under N_2 flow condition, followed by oxidative removal of the residual carbon reductant. Figure 3 shows SEM micrographs of the sample A calcined at 1450°C for 2 h. Macro-pores were found to be formed uniformly on the surface (Fig.3 (a)) and inside (Fig.3 (b)) of the bodies. Similar pores were seen in the sample B with the compositional adjustment.

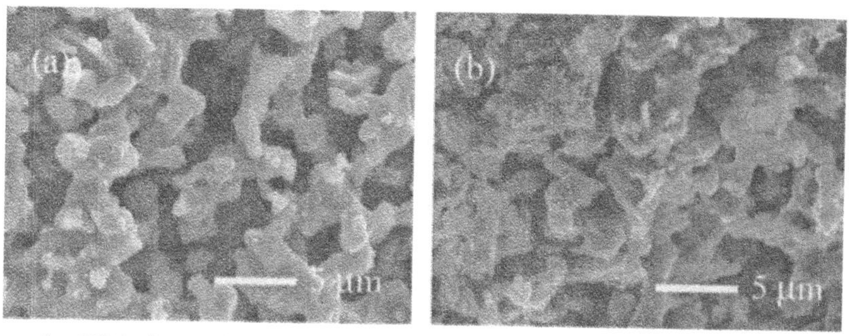

Figure 3. SEM micrographs of the sample A calcined at 1450°C for 2 h; (a) the surface and (b) fracture surface of the sample.

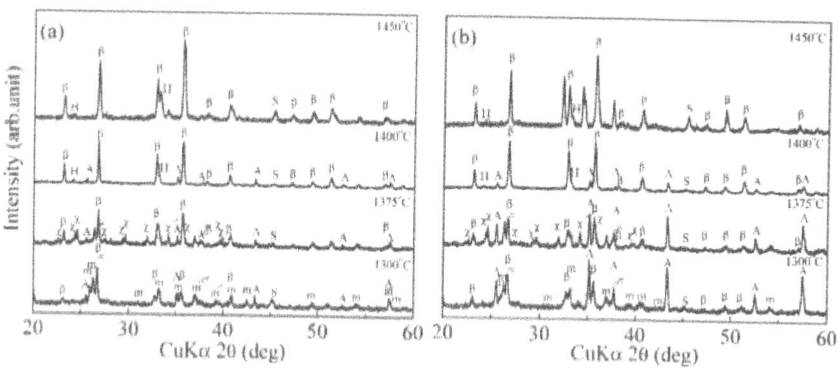

Figure 4. XRD patterns of the sample A (a) and B (b) calcined for 2 h at various temperatures. Symbols: q: quartz, m: mullite, H: Fe_2O_3, S: Fe_3Si, A: Al_2O_3, χ: $Si_3Al_6O_{12}N_2$, •: Si_2N_2O, β: $Si_3Al_3O_3N_5$, □: $SiAl_4O_2N_4$.

Figure 4 shows XRD patterns of the samples A and B calcined for 2 h at various temperatures. Increase of the calcination temperature results in an increase of β-sialon phase and a decrease of secondary phases. For the samples A and B calcined at 1300°C, β-sialon phase appeared, where the mullite phase present in the coal fly ash was still observed, but the quartz phase disappeared. On the sample A without Al_2O_3 addition for the compositional adjustment, α-Al_2O_3 peaks appeared at 1300°C. This is mainly because the mullite phase was decomposed under carbon coexisting and N_2 flow conditions. This result suggests that the β-sialon phase firstly formed from the quartz and the glass phase in the coal fly ash. When the calcination temperature was raised to 1375°C, the decomposition of the mullite phase was completed and secondary phases corresponding to χ-sialon ($Si_3Al_6O_{12}N_2$) and Si_2N_2O were formed. The secondary phases of χ-sialon, Si_2N_2O and α-Al_2O_3 completely disappeared at a calcination temperature above 1450°C for the sample A. Although most trends of the phase formation sequence for the sample B was agreed with that for the sample A, $SiAl_4O_2N_4$ was formed in the sample B calcined at 1450°C in addition to the β-sialon phase. This phase formation seems to come from the addition of Al_2O_3 for the compositional adjustment, which gives the smaller molar ratio of silicon/aluminum in the sample B compared to that of the sample A. In addition, silicon in the coal fly ash is considered to vaporize during the calcination.

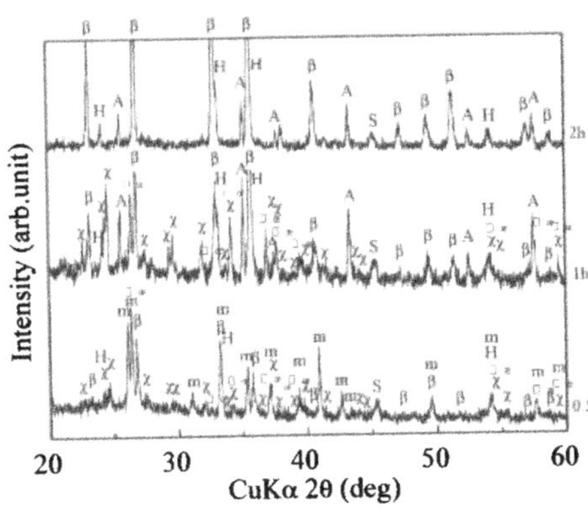

Figure 5. XRD patterns of the sample A calcined at 1400°C for various treatment times. Symbols; q: quartz, m: mullite, H: Fe_2O_3, S: Fe_3Si, A:Al_2O_3, χ:$Si_3Al_6O_{12}N_2$, •: Si_2N_2O, β: $Si_3Al_3O_3N_5$.

Although the coal fly ash contains minor components, such as sodium, potassium, calcium, manganese, titanium and iron, only the iron was detected as forms of hematite (Fe_2O_3) and iron silicide (Fe_3Si) by XRD measurement. Elemental distribution analysis by EDS demonstrated that the titanium and iron exited locally as spherical particles, on which silicon-based needle-like particles formed. Impurities of these elements worked as a catalyst for sialon formation.[4] On the other hand, other elements such as calcium and manganese were distributed uniformly on the matrix. Chemical analysis of the sample A before and after the calcination at 1400°C by ICP-AES showed that no loss of the minor elements could be seen except for alkaline elements.

Figure 5 shows XRD patterns of the sample A calcined at 1400°C for various treatment times. X-ray diffraction analysis showed that the quartz and mullite phases, which were originally contained in the coal fly ash, firstly began to decompose into intermediate phases such as χ-sialon, Si_2N_2O and Al_2O_3, and simultaneously β-sialon phase was formed through the intermediate phases during firing.

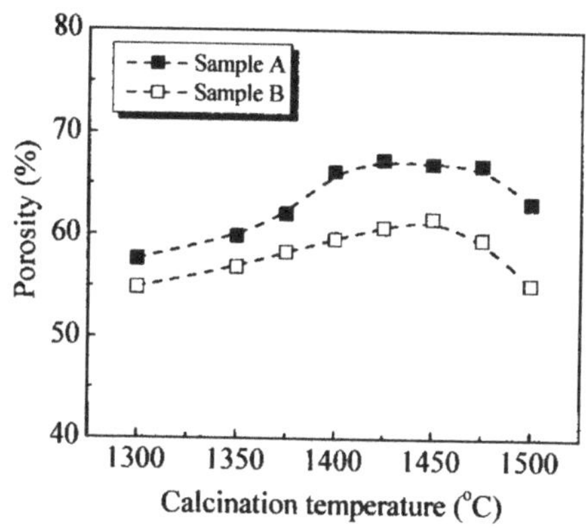

Figure 6. Dependence of porosity in the sample A and B on the calcination temperature.

Figure 6 shows the dependence of porosity in the samples A and B on the calcination temperature. Porous materials with the porosity of around 55-65% were obtained, which had a larger porosity compared to the calculated values shown in Table 3. The porosity tended to increase with increasing calcination temperature in a temperature range from 1300 to 1450°C. These are mostly attributed to the vaporization of silicon from the samples, since the weight loss of the powder compacts before and after the treatment gradually increased with increasing calcination temperature. On the other hand, further increase of the calcination temperature led to a decrease of the porosity, which was due to the densification of the samples. Figure 7 shows the pore size distribution of the sample A and B calcined at 1400°C for 2 h. Sample B was found to have a smaller pore size in comparison with that of sample A, because the sample B had fine Al_2O_3 particles for the adjustment, which led to the larger shrinkage of the samples in size; for example around 9% for the sample B, while 4% for the sample A at 1500°C.

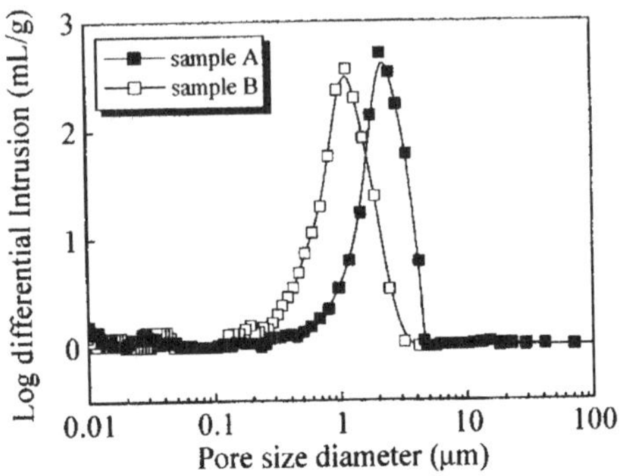

Figure 7. Pore size distribution of the sample A and B calcined at 1400°C for 2 h.

CONCLUSIONS

Sialon-based porous materials were prepared from the coal fly ash using carbothermal reduction and nitridation method. A nearly pure β-sialon phase could be obtained by the calcination of the powder compact under N_2 flow condition. The coal fly ash containing a large amount of amorphous phase was a good starting material for the preparation of the sialons. The sialon-based porous materials had the porosity of around 55-65% with a pore size distribution from sub-micrometer to several micrometers.

REFERENCES

[1]F.L. Riley, "Silicon Nitride and Related Materials," *J. Am. Ceram. Soc.*, **83**[2], 245–65 (2000).

[2]X.-J. Liu, X.W. Sun, J.J. Zhang, X.P. Pu, Q.M. Ge, L.P. Huang, Fabrication of β-sialon powder from kaolin," *Mater. Res. Bull.*, **38**, 1939-1948(2003).

[3]A.D. Mazzoni, E.F. Aglietti, E. Pereira, "β'-sialon preparation from kaolinitic clays," *Appl. Clay Sci.*, 7, 407–420 (1993).

[4]M. Vlasova, V.B. Vinokurov, O.N. Grigor'ev, A.D. Panasyuk, N.D. Bega, M. Kakazey, J.G. Gonzalez-Rodriguez, G. Dominguez-Patiño, M. Dominguez-Patiño, "Features of SiAlON synthesis from kaolin," *Mater. Sci. Eng.*, **A366**, 325–331 (2004).

Eco-Materials Processing and Design
Edited by Koji Watari and Soo-Wohn Lee

AQUEOUS TAPE CASTING OF LZSA GLASS CERAMICS

C.M. Gomes, F.N. Biscaia, J.T. Quinaud, O.R.K. Montedo, A.P.N. Oliveira, D. Hotza
Graduate Program in Materials Science and Engineering – PGMAT
Chemical Engineering Department – EQA
Federal University of Santa Catarina, Florianópolis, SC, Brazil, 88040-900

ABSTRACT

The effect of sodium alginate, as binder, and polyethyleneglicol, as plasticizer, on rheological and mechanical properties of suspensions used in water-based tape casting of LZSA glass ceramics was studied. The investigation was carried out using a statistical design approach. An aqueous suspension of a parent glass from the LZSA (Li_2O-ZrO_2-SiO_2-Al_2O_3) system was used. Viscosity, thixotropy and yield stress behavior were measured for the aqueous suspensions as a function of additive composition. Tensile strength of cast tapes was also measured. The experiments were shown to be statistically significant.

INTRODUCTION

The tape casting process is an economical method to produce thin, flat ceramic components[1,2], which has been applied to a wide range of materials and additives compositions. However, with the increasing of environmental problems, more processes commonly named "clean" are been required, in which no organic solvents are employed.

A great effort has been devoted by ceramists to develop water-based tape casting. However, this process has also a few drawbacks that may restrict its industrial application[3]. Compared to the organic-based process, aqueous systems have smaller tolerance to minor changes in processing parameters such as casting composition, drying conditions or film thickness[4]. Besides it is also important the choice of the additive settings that fulfill the requirements of water-based process as well as the environmental ones.

Alginate is a family of unbranched binary copolymers of 1,4-β-D-mannuronic acid and -L-guluronic acid extracted from marine brown algae with a wide variety of compositions and structures[5]. Alginate salts of alkali metals are soluble in hot or cold water, but in the presence of some cations, mainly Ca^{2+}, or in pH<3 they could become gel at room temperature. In this way, the choice for the ceramic material to be used with this additive is important, once the powder should not present impurities of gelation cations or turn suspension into acid media.

Glass ceramics belonging to the Li_2O-ZrO_2-SiO_2-Al_2O_3 (LZSA) system have been produced by different techniques in the last few years[6,7,8]. This system presents some interesting properties such as good chemical and thermal shock resistances and a very low thermal expansion coefficient (4 to $6x10^{-6}$ °C^{-1})[7]. Cordierite based glass ceramics have also been processed by tape casting, with the purpose of obtaining products with good dielectric and thermal properties[9].

The aim of this work is to study the viability to produce thin and flat tapes of glass ceramics from the Li_2O-ZrO_2-SiO_2-Al_2O_3 (LZSA) system by water-based tape casting using sodium alginate as binder.

EXPERIMENTAL PROCEDURE

As raw material, a parent glass powder from the LZSA system was used, with a mean particle size of 2.3 μm. Some residual crystalline phase was found, as can be seen in Figure 1(a), probably due to some crystallization process occurred during the melting of the starting oxides. The particles exhibit irregular and different shapes (Figure 1(b)).

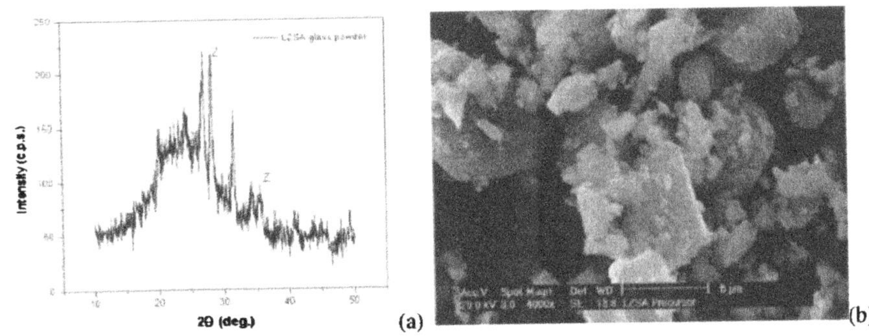

Figure 1. (a) XRD from LZSA glass powder, Z – zirconium silicate ($ZrSiO_4$); (b) morphology of parent glass LZSA powder, obtained by scanning electron microscopy (SEM).

The suspensions were prepared using sodium alginate (Na-alginate; Vetec, 2 wt.% solution, viscosity of 9.38 Pa.s) as binder, and polyethyleneglicol (PEG; Synth, molecular weight 400 g/mol) as plasticizer. A 3^2 factorial design was applied to formulate compositions from sodium alginate/PEG system (Statistica, StatSoft., 2000), with 3 repetitions at the central point. Table I shows the 12 statistical designed runs of the respective experiments.

Table I. Weight compositions of the suspensions.

Run	Glass powder (wt.%)	Water (wt.%)	Na-alginate (wt.%)	PEG (wt.%)
1	56.1	37.4	2.8	3.7
2	55.0	36.7	2.8	5.5
3	54.1	36.0	2.7	7.2
4	55.6	37.0	3.7	3.7
5	54.5	36.4	3.6	5.5
6	53.6	35.7	3.6	7.1
7	55.0	36.7	4.6	3.7
8	54.1	36.0	4.5	5.4
9	53.1	35.4	4.4	7.1
10	54.5	36.4	3.6	5.5
11	54.5	36.4	3.6	5.5
12	54.5	36.4	3.6	5.5

Each run was prepared in three steps: a) dispersion of the powder in water during 5 min in a planetary mill, to obtain a homogeneous mixture, b) binder addition and c) plasticizer addition, both with a mixture time of 10 min. The rheological characterization of the suspensions was performed with a rotational viscometer (Haake, VT550), at 25°C, and shear rates from 0.01 to 10 s^{-1}.

The green tapes were prepared by casting the as-prepared suspension on a flat surface of cellulose acetate with a manual doctor blade with a gap of 100 µm. All tapes were dried in air at room temperature for 24h.

Green densities were determined by measuring the weight and the apparent volume of samples with 2 x 2 cm^2 area. The thickness was measured by optical microscopy (Leica, DM400M). Tensile strength of the green tapes was also measured in a universal testing device (Emic, DL 2000), at a constant rate of 0.3 mm/min. The results are an average of, at least, 5 samples.

Tapes were sintered in an air oven (DL Eletric., Thermolab) according to the following thermal cycle: from room temperature until 300°C with a heating rate of 1°C/min., and from 300°C to 900°C with a heating rate of 10°C/min. The system was inertially cooled. The microstructural evolution of green and sintered tapes was observed using SEM (Phillips XL 30).

RESULTS AND DISCUSSION

Rheological Characterization

Table II shows the viscosity, yield stress, thixotropy and the shear thinning constant, n, measured for the suspensions of LZSA parent glass powder (Table I), as well as the coefficient of correlation, R^2, after fitting using the Herschel-Bulkley model[10].

Table II. Viscosity, yield stress, thixotropy, n index and R^2 of the suspensions with different binder/plasticizer content.

Run	Viscosity (Pa.s)	Yield stress (Pa)	Thixotropy (Pa/s)	n	R^2
1	6.17	5.45	1.15	0.8385	0.999
2	6.32	5.75	3.77	0.8398	0.999
3	6.14	7.29	4.30	0.7876	0.999
4	21.61	11.60	16.55	0.7481	0.999
5	19.54	9.72	21.32	0.7740	0.999
6	13.61	7.34	9.76	0.7859	0.999
7	33.89	8.09	3.50	0.7567	0.999
8	34.29	9.21	9.94	0.7523	0.999
9	33.21	8.19	15.58	0.7158	0.999
10	12.51	4.52	8.67	0.8291	0.999
11	17.07	4.52	1.18	0.7923	0.999
12	12.22	3.88	15.10	0.8152	0.999

Figures 2 and 3 show the surface fitted to the experimental data illustrating the influence of binder/plasticizer content on viscosity, yield stress, thixotropy and n constant. An increase of viscosity as well as the pseudoplastic behavior and thixotropy is observed by increasing amount of Na-alginate. However, only Na-alginate has a significant effect, according to an analysis of variance (not shown here), regarding to a confidence level of 95%. Suspensions for tape casting should present low viscosity, enough to ensure a homogeneous mass flow under the blade during casting, as well as a pseudoplastic behavior and absence of thixotropy[1].

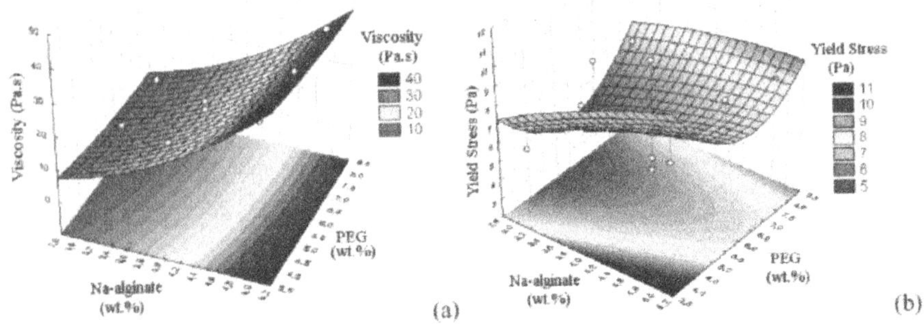

Figure 2. Response surfaces of (a) viscosity and (b) yield stress as a function of Na-alginate and PEG content.

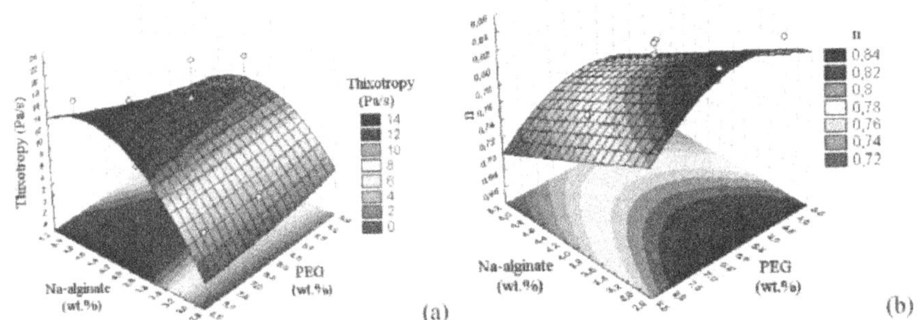

Figure 3. Response surfaces of (a) thixotropy and (b) n constant as a function of Na-alginate and PEG content.

To achieve the necessary requirements for tape casting, compositions with a low thixotropy and yield stress and a higher pseudoplastic behavior should be chosen[1]. In this way, compositions with Na-alginate contents higher than 4.4 wt.% in all range of PEG content can be used. Once rheology of the binders is critical for the rheological behavior of the final suspension and in a 2 wt.% Na-alginate solution a pseudoplastic behavior is highlighted, it was expected that higher contents of binders lead to an increase in the pseudoplasticity of the suspensions as well as in the thixotropy. Table III presents the results of tensile strength and green density of tapes cast from suspensions of LZSA parent glass powder (Table I) in aqueous media.

Table III. Tensile strength and green density of the 12 compositions studied.

Run	Tensile Strength (MPa)	Green density (g/cm³)
1	0.30 ± 0.05	1.41 ± 0.00
2	0.52 ± 0.08	1.45 ± 0.45
3	0.33 ± 0.16	1.26 ± 0.25
4	0.60 ± 0.10	1.46 ± 0.06
5	0.55 ± 0.07	1.68 ± 0.15
6	0.62 ± 0.07	1.17 ± 0.02
7	0.78 ± 0.12	1.25 ± 0.23
8	0.72 ± 0.05	1.54 ± 0.09
9	0.88 ± 0.14	1.37 ± 0.03
10	0.65 ± 0.03	1.38 ± 0.16
11	0.65 ± 0.03	1.41 ± 0.05
12	0.65 ± 0.03	1.41 ± 0.05

Figures 3 (a) and (b) show the influence of binder and plasticizer on the tensile strength and green density of the tapes.

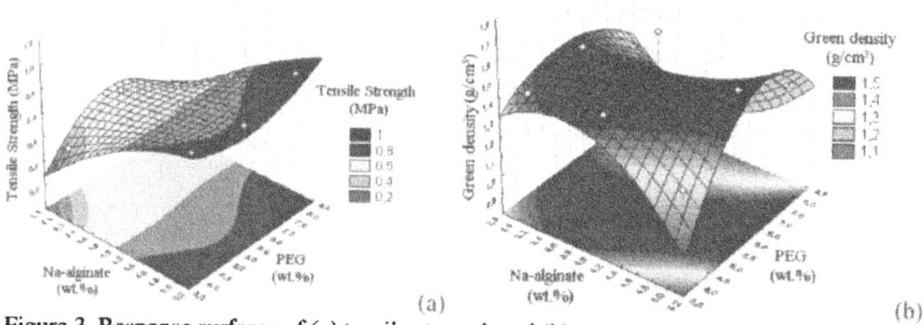

Figure 3. Response surfaces of (a) tensile strength and (b) green density as a function of the Na-alginate and PEG content.

Figure 3 (a) shows that higher values of tensile strength are reached for compositions with higher amounts of Na-alginate (> 4.4 wt.%), independent of the plasticizer content, i.e., the binder influences more significantly on the tensile strength. However, for the green density, the amount of PEG was crucial and reached a maximum at 5.5 wt.%. Figures 4 (a) and (b) show the green microstructure of compositions 7 and 9, respectively.

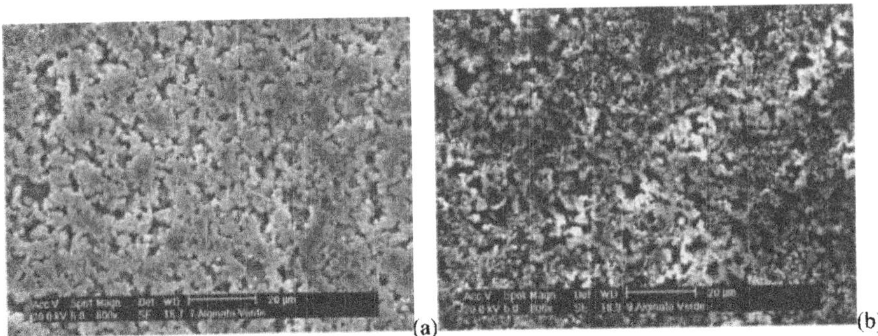

Figure 4. Green microstructure of cast tapes from compositions with: (a) 4.6 wt.% Na-alginate/ 3.7 wt.% PEG (run 7), and (b) 4.4 wt.% Na-alginate/ 7.1 wt.% PEG (run 9).

Although both compositions exhibit similar values of tensile strength, the green density of composition 9 was higher than the composition 7. This can be explained due to differences in the density of the original additives (Na-alginate = 1.012 to 1.022 g/cm^3, and PEG = 1.130 g/cm^3).

Figure 5 shows the X-ray diffractograms of LZSA glass ceramics formed. It can be seen the formation of the crystalline phases, characteristic of the LZSA glass ceramics.

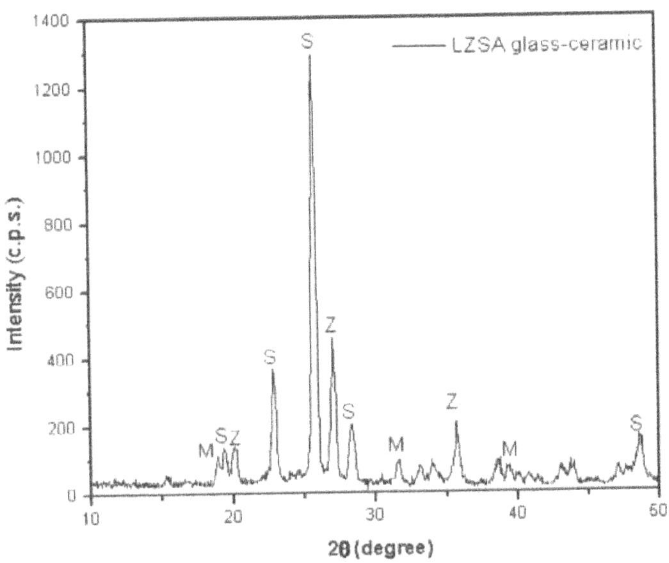

Figure 5. XRD of LZSA glass ceramics tapes, where S is β-spudomene- ($LiAlSi_3O_8$), Z is zirconium silicate ($ZrSiO_4$), and M is lithium metasilicate (Li_2SiO_3).

Figure 6 (a) and (b) shows the microstructure sintered of the tapes from compositions 1 and 9, respectively.

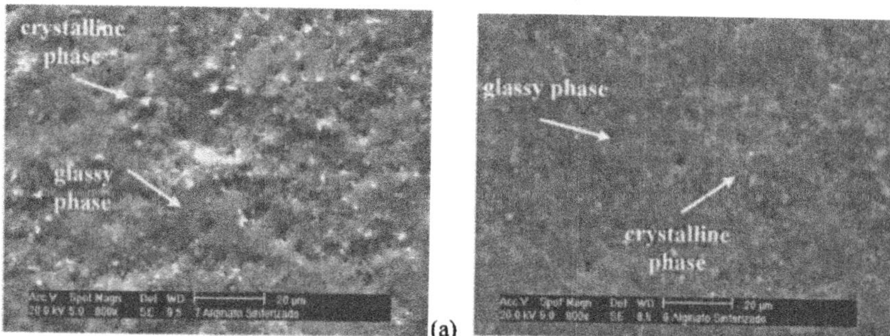

(a) (b)

Figure 6. Microstructure sintered of the tapes from compositions with: (a) 4.6 wt.% Na-alginate/ 3.7 wt.% PEG (run 7), and (b) 4.4 wt.% Na-alginate/ 7.1 wt.% PEG (run 9).

Both compositions exhibit similar microstructures, with a residual glassy phase and some crystalline phases dispersed in the glass matrix, as indicated by the arrows. As already observed by Bitterlich et al.[11] for YSZ slurries, the green density of the tapes has no visible effect on the final density of the sintered tapes. All sintered tapes exhibit approximately the same density, although the green densities varied substantially with different binder contents.

CONCLUSIONS

LZSA thin tapes were obtained from a water-based tape casting process. The additives system (Na-alginate/PEG) was viable to be employed for this process. All compositions presented a pseudoplastic behavior with no visible thixotropy and yield stress. Compositions with higher amounts of Na-alginate presented higher values of tensile strength.

The sintered microstructures formed were similar and independent from the additive content.

ACKNOWLEDGEMENTS

The authors are very thankful for the financial support from CAPES/Brazil.

REFERENCES

[1] R.E. Mistler, and E.R. Twiname, "Tape Casting – Theory and Practice", *The American Ceramic Society*, Ohio (2000).

[2] A. Roosen, "Basic Requirements for Tape Casting of Ceramic Powders", *Ceram. Trans.* **1**, 675-692 (1998).

[3] I. Santacruz, A.C. Gutiérrez, M.I. Nieto, and R. Moreno, "Application of Alginate Gelation to Aqueous Tape Casting Technology", *Mat. Resear. Bull.*, **37**, 671-682 (2002).

[4] P. Nahass, W.E. Rhine, R.L. Pober, H.K. Bowen, and W.L. Robbins, "A Comparison of Aqueous and Non-Aqueous Slurries for Tape Casting and Dimensional Stability in Green Tapes", *Ceram. Trans.*, **15**, 355 – 64 (1990).

[5] S.T. Moe, K. I. Draget, G. Skjak-Braek, O. Smidsrod, in: A. M. Stephen, Food Polysaccharides and their Applications, Marcel Dekker, New York, p. 245, (1995).
[6] L. Giassi, O.R.K. Montedo, M.C. Fredel, A.P.N. Oliveira, "Caracterização, Processamento e Propriedades de Vitrocerâmicos do Sistema LZSA Obtidos via Moldagem por Injeção", in Proceedings of 48° Congresso Brasileiro de Cerâmica, Curitiba, PR, Brazil, (2004).
[7] O.R.K. Montedo, A.P. Novaes de Oliveira, A.N. Klein, PTECH 2001 – 3rd Latin-American Conf. on Powder Technology, Florianópolis, SC, Brazil, November (2001).
[8] E. Sousa, C.B. Silveira, T. Fey, P. Greil, D. Hotza, A.P.N. Oliveira, "LZSA Glass Ceramic Foams Prepared by Replication Process", *Advances In Applied Ceramics*, **104**, 22-9 (2005).
[9] S. Mei, J. Yang, X. Xu, S. Quaresma, S. Agathopoulos, J.M.F. Ferreira, "Aqueous Tape Casting Processing of low Dieletric Constant Cordierite-Based Glass-Ceramics – Selection of Binder", *J. Eur. Ceram. Soc.*, in Press (2004).
[10] J. S. Reed, Principles of Ceramic Processing. Wiley-Interscience, New York, 277-309 (1995).
[11] B. Bitterlich, C. Lutz, A. Roosen, "Rheological Characterization of Water-Based Slurries for the Tape Casting Process", *Ceram. Intern.*, **28**, 675 – 83 (2002).

Eco-Materials Processing and Design
Edited by Koji Watari and Soo-Wohn Lee

DIRECT COAGULATION CASTING OF AQUEOUS SILICON CARBIDE SLURRY BY ADDITION OF BORON NITRIDE

Seizo Obata
Gifu Prefectural Ceramics Research Institute
3-11 Hoshigadai
Tajimi 507-0811, Japan

Naoyuki Kato, Osamu Sakurada and Minoru Hashiba
Faculty of Engineering, Gifu University
1-1 Yanagido
Gifu 501-1193, Japan

ABSTRACT

A new method for direct casting of aqueous silicon carbide (SiC) slurries with carbon and boron carbide as sintering aids has been developed. The SiC used was beta-SiC, and carbon and boron carbide were used as sintering aids. High solid-loaded SiC slurries were obtained by addition of ammonium polystyrene sulfonate and tetramethyl ammonium hydroxide at pH 11, and the thickening limit of the slurry with good fluidity was found at a solid content of 72 wt%. When hexagonal boron nitride (BN) was added to the SiC slurries, the ionic strength of the slurry increased gradually by elution of BN. The cast slurries were coagulated, changing their properties from a viscous liquid to a homogeneous, rigid, green body in the non-porous mold because of gradual elution of the BN particles added just before casting. The high green density of 63% theoretical density (T. D.) was obtained by addition of 0.1–1.1 wt% BN. However, the addition of ~1.1 wt% BN to the SiC slurry with carbon did not fully promote densification of the sintered SiC body. With further addition of B_4C as a sintering aid to the SiC slurries with carbon, the density of the sintered body reached 97% T. D. using this new direct casting method.

INTRODUCTION

Silicon carbide (SiC) materials are prime candidates for structural ceramic parts, such as gas turbine engines, heat exchangers, wear-resistant seals, and diffusion furnace parts for semiconductor manufacture [1-3]. For fabrication of these SiC components with high uniformity, high performance, and high reliability, it is important to minimize defects, such as hard agglomerates, pores, and cracks. The forming process for these materials as well as the preceding SiC powder preparation process is very important in the fabrication of ceramic parts as defects introduced during these processes will normally remain in the product even after successful sintering. Shaping of SiC through aqueous media [4, 5] is especially important in terms of environmental impact, safety, and cost.

Colloidal processing is a promising method to produce complex-shaped parts with a reduction in the number and size of defects and increased reliability [6, 7]. This colloidal process is suitable for fabricating near net shaped components for which little machining is required. In this method, enhancement of the dispersion and fluidity of the SiC slurry is a key technique for fabricating structural SiC parts.

In previous studies, we showed that SiC bodies with high green and sintered density could be obtained through the slip casting process using aqueous highly solid-loaded slurries with sintering aids and dispersants, such as anionic polymer dispersant and tetramethyl ammonium hydroxide (TMAOH) at pH ~10–11 [8-10]. The flow curves and flow point data indicated that the dispersion and fluidity of SiC slurries were effectively enhanced by addition of TMAOH, as an oxide film was formed on the surface of SiC particles and the SiC particles behaved like SiO_2 particles in their colloidal stability; the absolute zeta potentials of the SiC particles showed large values in the basic region [11, 12].

Usually, boron carbide and carbon powders are used as sintering aids [13]. Boron carbide had no effect on the fluidity of aqueous SiC slurries [8]. Enhancement of the dispersion and fluidity of carbon particles in the slurry was a key subject for slip casting of aqueous slurries, because the carbon powder has a hydrophobic surface and high specific surface area. We reported that dispersion and fluidity of aqueous carbon slurry were enhanced by addition of a polyelectrolyte with both hydrophobic and hydrophilic groups to improve the surface properties of the carbon particles from hydrophobic to hydrophilic [14].

Recently, new forming processes based on colloidal processing [15] have been developed for near net shaping of complex ceramic parts, such as gel casting [16] and direct coagulation casting (DCC) [17, 18]. DCC is a novel near net shaping technique for complex ceramic parts with homogeneous microstructure and high green density. This process is based on the casting of highly solid-loaded slurry into a mold and then using a time-delayed chemical reaction in the slurry, either by shifting the pH toward the isoelectric point (IEP) of the slurry or by increasing the electrolyte concentration in the slurry [15]. Both reaction paths can be used to minimize the repulsive force between the particles, which transforms to a stable slurry from the liquid state to a stiff particle network. DCC is superior to slip casting in several respects: (i) shaping can be performed without any restriction of the mold; and (ii) more complex ceramic parts with homogeneous microstructure and high green density can be fabricated. We reported that the Al_2O_3 slurry was solidified by increasing the ionic strength of the slurry by the addition of aluminum nitride or yttrium oxide[19, 20]. Aluminum nitride and/or yttrium oxide form various aluminum and/or yttrium species by hydrolysis with changing pH: *i.e.*, ions, oxide ions, hydroxide, and ammonium ions. The electrostatic repulsive force between the Al_2O_3 particles was markedly reduced by the addition of these compounds. We noted that boron nitride (BN) also retained the action of solidification because this was gradually eluted into aqueous media and we expected that the addition of BN should be available for densification of sintered SiC bodies as a sintering aid because of the homogeneous dispersion of eluted boron species in the green body.

The present study was performed to elucidate the effects of BN on the DCC process of aqueous high solid-loaded SiC slurries with sintering aids, which were dispersed and thickened using TMAOH and polyelectrolyte. We measured the following processing factors: (i) the elution rate of BN into aqueous media; (ii) the onset time of solidification of the slurry with various amounts of BN added; and (iii) the densities of green and sintered bodies.

EXPERIMENTAL

Materials

The silicon carbide used in this study was beta-SiC (UF, Ibiden Co., Ltd., Gifu, Japan). Carbon black powder (#2650, Mitsubishi Chemical Co., Tokyo, Japan) and boron carbide (B_4C: Denka Boron #1200, Denki Kagaku Kogyo Co., Tokyo, Japan) were used as the boron and carbon sources for sintering aids. Commercial hexagonal BN powder (h-BN, UHP-1S, Showa Denko K. K., Tokyo, Japan) was used to control the ionic strength of the slurry. Table 1 shows

Table 1 Characteristics of the powders

	SiC UF	Carbon #2650	B_4C #1200	BN UHP-1S
Mean particle size (μm)	0.29	0.013	1.2	1.0
Specific surface area (m^2/g)	19.9	320	18-20	24

the characteristics of the powders supplied by the manufacturer. Ammonium polystyrene sulfonate (PSS; SN-Dispersant 5047, San Nopco Ltd., Tokyo, Japan) and tetramethyl ammonium hydroxide (TMAOH) were selected as dispersing and fluidizing agents. The pH of the slurries was adjusted with TMAOH and HNO_3.

Shaping processing and sintering

Slurries containing water, PSS, silicon carbide, carbon, and boron carbide, which were added to the slurries in fixed amounts on a dry-weight basis (dwb) for the silicon carbide or carbon powders, were prepared by ball milling using SiC balls for 24 h at room temperature. The amounts of carbon powder and PSS added were 1 wt% on dwb for SiC and 15.5 wt% on dwb for carbon, respectively. The water used to prepare all of the slurries was distilled and purified using a Milli-Q system (Milli-Q Plus, Millipore Co., Bedford, MA, USA).

After ball milling, an appropriate amount of BN powder was added gradually to the slurries and mixed using a planetary mixer (MS-SNB-350B, Matsuo Sangyo Co., Ltd., Osaka, Japan) for 150 s. The slurries were degassed under vacuum and cast into nonporous polypropylene molds. The slurries were left to stand in a climatic chamber with controlled humidity (97%) and temperature (27°C). The solidified bodies were removed from the mold and dried at room temperature. Dried green bodies (80 × 50 × 6 mm) were subjected to pressureless sintering in an Ar atmosphere at 2150°C for 1 h. The slip casting using a gypsum mold was carried out a control experiment.

Measurements

To determine the amount of BN eluted, 1 g of BN was added to 50 mL of the aqueous solution with/without various concentrations of PSS at pH 11 and left to stand for various periods. Boron concentration in the supernatant of the suspension was determined using an ICP-AES (PS1000UV, Leeman Labs Inc., NH, USA). Sedimentation volume were measured after pouring 100 ml of each 2 vol% BN slurry into graduated cylinders, which were then sealed and left to stand for 48 h prior to measurement.

The rheological behaviors of the slurries were measured at 25°C using a controlled-stress rheometer (RS-50, Haake, Karlsruhe, Germany) with a parallel plate sensor 35 or 60 mm in diameter. The onset time of solidification of the slurry was determined by measurement of the storage modulus (G') at 25°C using a controlled stress rheometer with a double gap cylinder (DG-41). In this study, the time at which G' began to increase was determined as the solidification onset time of the slurry. The electrokinetic behavior was determined for 5 vol% slurries using an acoustophoretic spectrometer (DT-1200, Dispersion Technology Inc., NY, USA) at 25°C. The densities of the green compacts were calculated from their size and weight. The fired densities were measured by the Archimedes method. The microstructure of sintered

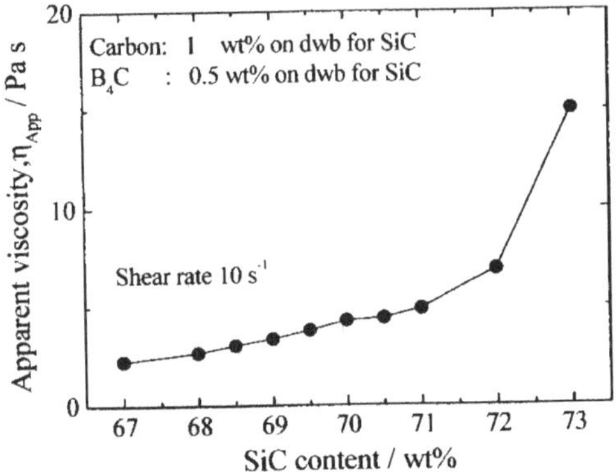

Fig. 1 Apparent viscosities of SiC slurries with carbon and B_4C as a function of solid content at pH 11.

bodies was observed by scanning electron microscopy (SEM, S-4300, Hitachi High-Technology Co., Tokyo, Japan).

RESULTS AND DISCUSSION

Thickening of the SiC slurry with sintering aids

A thickened slurry with good fluidity was preferable for colloidal processing. The fluidity of electrostatically stabilized SiC slurry was dependent on solid content and powder characteristics, such as particle size, size distribution, and aspect ratio. Previously, we reported that an aqueous high solid-loaded SiC slurry with good fluidity can be prepared by addition of an anionic polymer dispersant and TMAOH at pH ~10–11 [8-10]. In the present study, the SiC slurries were prepared at pH 11. Figure 1 shows the apparent viscosities of SiC slurries with carbon and B_4C as a function of solid loading at pH 11. Carbon and B_4C were added to 1 wt% and 0.5 wt% on dwb for SiC, respectively. The apparent viscosity increased gradually up to 72 wt% SiC and then increased abruptly. We determined that the thickening limit of the slurry with good fluidity, which was preferable to the DCC method, was at a solid content of 72 wt%.

Solidification behavior of slurries

Figure 2 shows the effects of amount of BN added to 72 wt% SiC slurries on the solidification onset time. The onset time decreased markedly up to the addition of 0.7 wt% BN and showed a slight decrease at 0.7–1.1 wt%. Solidification of SiC slurry did not start without BN. On the other hand, with addition of BN above 1.1 wt%, the slurries began to solidify during mixing with a planetary mixer. Therefore, the onset time could not be determined and the slurries could not be cast into the mold. The optimum amount of BN was found to lie within the range of 0.1 to 1.1 wt% in the dwb to SiC powder for DCC. The elution behavior of BN into aqueous media was the most important parameter in the DCC method. Figure 3 shows the time dependence of the amount of BN eluted into the aqueous media in the presence of various amounts of PSS at pH 11. The amount of boron in aqueous media with/without PSS increased gradually with elution time and leveled off at 1 h. The amount of boron eluted into the media

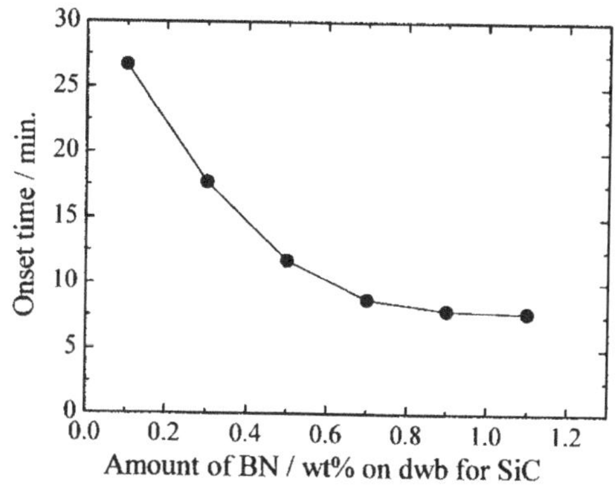

Fig. 2 Effects of amount of BN added to 72 wt% SiC slurries with sintering aids on the solidification onset time.

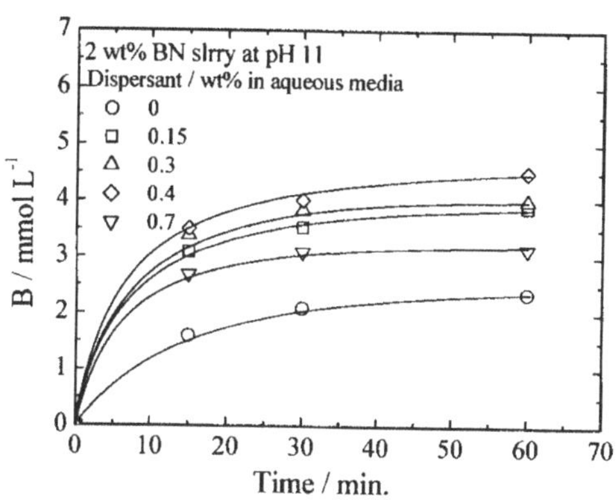

Fig. 3 Elution of BN into aqueous media with various concentrations of PSS at pH 11.

increased with addition of PSS in aqueous media until 0.4 wt% and then the amount of boron eluted decreased by addition of BN to 0.7 wt% PSS solution. To evaluate the effects of addition of PSS on elution of boron from BN powder in aqueous media, the sedimentation volumes of BN slurries with various amounts of PSS were measured as shown in Fig. 4 together with the amount eluted at 1 h. The sedimentation volume of BN left to stand for 48 h decreased with addition of increasing amounts of PSS in aqueous media and then showed the minimum volume by addition of 0.3 wt% PSS. Addition of excess PSS above 0.7 wt% increased the sedimentation volume due to the increase in ionic strength of the media. This observation indicated that dispersion of BN powder, which consists of plate-like particles and has a tendency to agglomerate, was enhanced

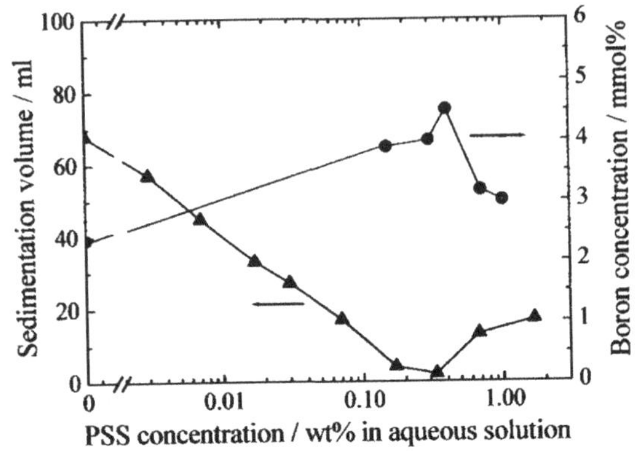

Fig. 4 Sedimentation volume of BN particles and eluted amount from BN as a function of PSS concentration in aqueous media.

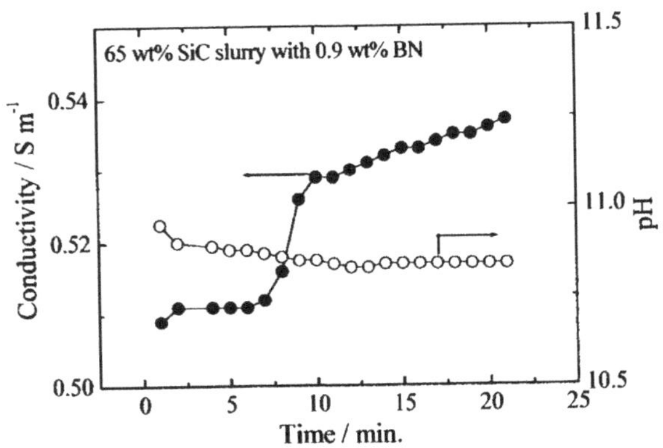

Fig. 5 Effects of elution time on conductivity and pH of 65 wt% SiC slurry containing sintering aids in the presence of 0.9 wt% BN.

by addition of PSS. The increase in elution of BN into the media with PSS could be explained by the increase in BN surface area due to deflocculation of BN particles in aqueous media. The elution of BN into the media was promoted by dispersion of BN particles affecting the dispersal of PSS.

We examined how the addition of BN affected solidification of SiC slurries. Figure 5 shows the effects of elution time of addition of 0.9 wt% BN on conductivity and pH of 65 wt% SiC slurry containing sintering aids. After addition of BN, conductivity of the slurry showed a constant value until 6 min, then increased markedly from 6 to 10 min, and then increased gradually after 10 min. The onset time of slurry coagulation by addition of 0.9 wt% BN was 8

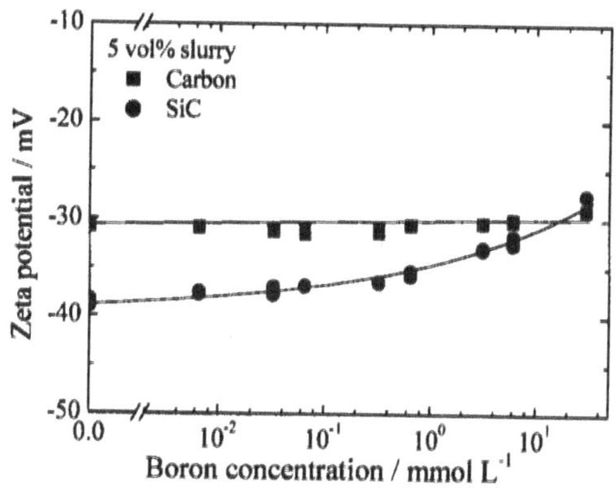

Fig. 6 Zeta potentials of SiC and carbon particles as a function of boron concentration.

min after addition of BN, as shown in Fig. 2. The pH value of the slurry shifted from 10.9 to 10.7 after addition of BN. This change in pH indicated a slight effect on coagulation of SiC slurries because aqueous high solid-loaded SiC slurry with good fluidity could be prepared at pH ~10–11. Figure 6 shows the zeta potential of SiC and carbon particles as a function of boron concentration. SiC and carbon slurries were prepared by addition of TMAOH and PSS as dispersants, respectively, at pH 11. Boron standard solution for atomic absorption spectrometry was used to control the boron concentration in the slurry. Zeta potential values of carbon particles were almost constant and were independent of the boron concentration. On the other hand, absolute values of those of SiC particles decreased gradually with increasing boron concentration. These results indicated that the dissolution of BN into the slurry gradually increased the ionic strength of the slurry, which should be mainly coagulated the stabilized slurry. The increase in ionic strength by elution from BN affected the coagulation of SiC particles in the slurry because of the reduction of electrostatic repulsive force between the particles.

Evaluation of green and sintered bodies

The green and sintered bodies were obtained by the DCC method by addition of BN using high solid-loaded SiC slurry with good fluidity. For comparison, green and sintered bodies were prepared by the slip casting process using aqueous SiC slurries containing BN and carbon. Figure 7 shows the densities of the green and sintered bodies formed *via* DCC and slip casting as a function of the amount of BN added. The green density values of the bodies formed through both DCC and slip casting were not different, and were 62–63% of the theoretical density of SiC (T. D.: 3.21 g/cm^3) due to the high degree of packing in the green body. The sintered density values of the bodies prepared *via* slip casting increased gradually with increasing amount of BN added, and the maximum value was 87.5% T. D. of the body with addition of 1.2 wt% BN. On the other hand, those of the bodies formed *via* DCC increased gradually with increasing amount of BN added at 0.3–1.1 wt% BN, and reached a maximum value of 93.7% of T. D. with addition of 1.1 wt% BN to the body. If the amount of BN added acted as a sintering aid for SiC, sintered

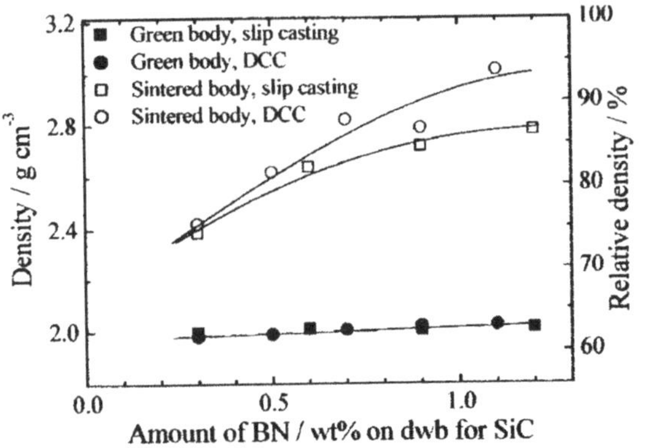

Fig. 7 Densities of green and sintered bodies formed *via* DCC and slip casting as a function of the amount of BN.

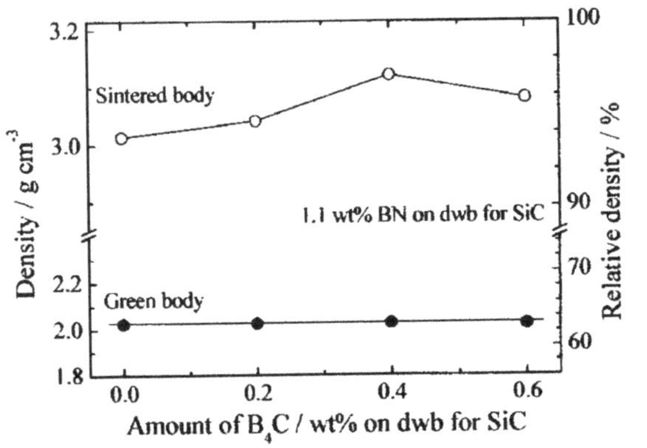

Fig. 8 Densities of green and sintered bodies formed *via* DCC as a function of the amount of B_4C.

bodies obtained through the two methods could be densified and would have shown almost the same sintered density values. However, the sintered densities of the bodies formed *via* slip casting were lower than those of bodies formed *via* the DCC process. In addition, with addition of increasing amounts of BN, the differences in the sintered density values of bodies formed *via* DCC and slip casting increased. Slip casting uses permeable molds, where media is drained out from the slurry, while DCC uses impermeable molds, where the media is not drained out because of time-delayed coagulation of the initially stabilized slurry. In the DCC process, boron species dissolved in the SiC slurry were retained in the solid body and dispersed in SiC grains and

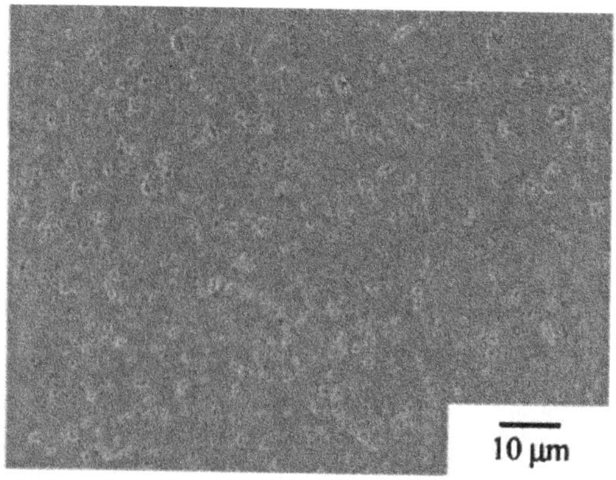

Fig. 9 Microstructure of the polished cross-section of a sintered body with 1 wt% carbon and 0.4 wt% B_4C *via* the DCC process by addition of 1.1 wt% BN.

precipitated between the particles. For this reason, boron species eluted from the BN surface should function as sintering aids for densification of SiC powders. However, dense sintered SiC bodies were not obtained by addition of 0.1–1.1 wt% BN.

For the preparation of SiC bodies with higher sintered density, it is necessary to add a boron component to the slurry. SiC bodies were formed through the DCC process by addition of 1.1 wt% BN using SiC slurries containing carbon and B_4C. Figure 8 shows the densities of the green and sintered bodies formed *via* DCC as a function of the amount of B_4C added. The green density value of bodies, 63% T. D., was not dependent on the amount of B_4C added. The sintered density value of the body increased gradually with increasing amount of B_4C added until 0.4 wt% on dwb for SiC, and the maximum value was 97.1% T. D. with the addition of 0.4 wt% B_4C. This sintered density of the silicon carbide body formed *via* DCC by addition of BN was equivalent to that of silicon carbide bodies with carbon and B_4C formed *via* the slip casting process. These higher sintered densities of SiC reflected the results obtained for the green bodies. Figure 9 shows the microstructure of the sintered body with 1wt% carbon and 0.4 wt% B_4C in the DCC process by addition of 1.1 wt% BN. In this result, the sample showed a dense microstructure and homogeneous pore distribution. Moreover, the largest pore size of this sample *via* DCC process was < 5 μm and very small. These results indicated that carbon and boron carbide as sintering aids and BN could be distributed homogeneously in silicon carbide slurries and green bodies obtained by the DCC process. Thus, BN in the DCC process should be suitable for fabrication of higher density green and sintered SiC bodies with a homogeneous structure.

CONCLUSIONS

In the new direct casting method developed in the present study, BN was effective as a starting reagent to induce solidification of the SiC slurry with carbon and B_4C stabilized with PSS and TMAOH. With addition of BN into the SiC slurry, the elution of BN gradually increased the ionic strength of the slurry. As a result, the thickened and fluidized SiC slurries containing carbon by addition of BN just before casting were coagulated, showing a change in properties from a viscous liquid to the final homogeneous, rigid, green body. A green body with high density (63% T. D.) was obtained by addition of 0.1–1.1 wt% BN. On the other hand, BN did not promote the sintering of SiC powder. With further addition of B_4C as a sintering aid to the SiC slurries with carbon, the sintered density of 97% T.D. was obtained using this new direct casting method.

REFERENCES

[1] M. Furukawa and S. Baba, "High Density Sintered Silicon-Carbide Parts"; pp. 409-15 in *SiC Ceramics*, Edited by S. Somiya and Y. Inomata. Uchida Rokakuho, Tokyo, Japan, 1988.

[2] H. Matsuo, *FC Report*, **10**. 382-89 (1992).

[3] Y. Furuse, *F C Report*, **11**. 235-39 (1993).

[4] J. N. F. Ferreira and H. M. M. Diz, "Effect of Slurry Structure on the Slip Casting of Silicon Carbide," *J. Eur. Ceram. Soc.*, **10**. 59-64 (1992).

[5] S. Sano, T. Banno, M. Maeda, K. Oda and H. Shibasaki, "Slip Casting and Sintering of Silicon Carbide: Slip Preparation of Silicon Carbide Powder Produced by Acheson Method," *J. Ceram. Soc. Jpn*, **104**. 984-88 (1996).

[6] F. F. Lange, "Powder Processing Science and Technology for Increased Reliability," *J. Am. Ceram. Soc.*, **72**. 3-15 (1989).

[7] J. A. Lewis, "Colloidal Processing of Ceramics," *J. Am. Ceram. Soc.*, **83.** 2341-59 (2000).

[8] S. Obata, K. Oda, H Kingugasa, H. Asano, O. Sakurada, M. Hashiba, K. Hiramatsu and Y. Nurishi, "Slip Casting of Basic Aqueous SiC Slurry Containing Styrene/Maleic Copolymer," *Key Eng. Mater.*, **159-160.** 157-162 (1999).

[9] S. Obata, H. Asakura, O. Sakurada, M. Hashiba and Y. Nurishi, "Effect of Free Carbon Contained in SiC Raw Powders on Slip Casting and Sintering Processes," *Key Eng. Mater.*, **206-213**. 365-368 (2001).

[10] S. Obata, H. Asano, O. Sakurada, M. Hashiba and Y. Nurishi, "Fabrication of Dense Silicon Carbide Through Aqeous Slurries Containing Well-dispersed Carbon as a Sintering Aid," *J. Mater. Sci.*, **40**. 757-760 (2005).

[11] J. Binner and Y. Zhang, "Characterization of Silicon Carbide and Silicon Powders by XPS and Zeta Potential measurement," *J. Mater. Sci. Lett.*, **20**. 123-126 (2001).

[12] M. N. Rahaman, Y. Boiteux and L. C. Dejongle, "Surface Characterization of Silicon Nitride and Silicon Carbide," *J. Am. Ceram. Soc. Bull.*, **65**. 171-6 (1986).

[13] S. Prochazka, "The Role of Boron and Carbon in The Sintering of Silicon Carbide"; pp. 171-81 in Special Ceramics 6, Edited by P. Popper, British Ceramic Research Association, Stoke-on-Trent, England, 1975.

[14] S. Obata, H. Asakura, H. Asano, O. Sakurada, M. Hashiba, and T. Shimada, "Enhancement of Dispersion and Fluidity of Aqueous Carbon Slurry by Addition of Various Dispersants," *J. Ceram. Soc. Jpn.*, **112**. S187-90 (2004).

[15] W. M. Sigmund, N. S. Bell and L. Bergstrom, "Novel Powder-Processing Methods for Advanced Ceramics," *J. Am. Ceram. Soc.*, **83.** 1557-74 (2000).

[16] L. Zhou, Y. Huang and Z. Xie, "Gelcasting of Concentrated Aqueous Silicon Carbide Suspension," *J. Eur. Ceram. Soc.*, **20.** 85-90 (2000).

[17] W. Si, T. J. Graule, F. H. Baader and L. J. Gauckler, "Direct Coagulation Casting of Silicon Carbide Components," *J. Am. Ceram. Soc.*, **82.** 1129-36 (1999).

[18] W. Li, H. Zhang, Y. Jin and M. Gu, "Rapid Coagulation of Silicon Carbide Slurry via Direct Coagulation Casting," *Ceram. Int.*, **30.** 411-16 (2004).
[19] M. Hashiba, M. kawamoto and O. Sakurada, "Shaping of Alumina Body through an Eco-conscious Process: Direct Casting of Alumina Slurry by Hydrolysis of Aluminum Nitride," *Trans. Mater. Res. Soc. Jpn.*, **29**. 2041-44 (2004).
[20] O. Sakurada, S. Imaeda and M. Hashiba, "Direct Casting of Aqueous Alumina Slurries Using Increase of Ionic Strength by addition of Yttria," *Trans. Mater. Res. Soc. Jpn.*, **29**. 2045-48 (2004).

Eco-Materials Processing and Design
Edited by Koji Watari and Soo-Wohn Lee

EFFECT OF AQUEOUS BASED FATTY ACID CHAIN LENGTH ON THE THERMAL DECOMPOSITION AND RHEOLOGICAL PROPERTIES OF EXTRUDED THIN WALLED PASTES

C.R. August, R.A. Haber, L.E. Reynolds
Rutgers University, Department of Ceramic & Materials Engineering
607 Taylor Rd
Piscataway, NJ 08854

ABSTRACT

Extrusion of environmental catalysts used in automotive applications has trended to increasingly thinner walls (i.e. 0.015 – 0.002 in.). This trend has spawned increasing interest in knowledge of the interactions within the bulk of an extrudable body. A combination of bulk and surface contributions must be understood and both extrusion pressures and green stiffness must be controlled. Approaches to improve these factors have included the use of aqueous solvents blended with lubricants such as oleic acid, linoleic acid, ricinoleic acid and combinations of these derivatives. However, problems associated with this solvent removal and burn out have limited the widespread use of these systems. In this study, a series of fatty acids ranging from C_{10} – C_{18} (caprylic to oleic) have been incorporated into an aqueous based extrusion body. Specific adsorption to particle surface will be shown. In addition, the thermal decomposition properties of these lubricants will be outlined. The effect of fatty acid adsorption to an oxide based formulation will be demonstrated. Torque rheometry, capillary rheometry and mechanical compliance will be shown.

INTRODUCTION

As manufacturing requirements for thin walled ceramics become more and more demanding, it is important to gain an understanding of the interactions both between organic additives and inorganic compounds and between inorganic compounds of differing surface characteristics. Optimization of chemistry and quantity of organic additives to an extrusion batch can only be achieved through a greater understanding of the interactions between components in that batch. In this exploration, fatty acid salts ranging from sodium stearate to sodium caprylate will be used as additives in single and dual component pastes. These pastes will be constructed using alumina and talc, two inorganics with very different surface charges. The charge behavior of these inorganics will be looked at, along with the effect of the additives on the paste rheology. The thermal behavior of the fatty acid salts will also be examined.

Polymeric dispersants are widely used in manufacturing. It has been found that these dispersants can act in two methods. Adsorbed polymers can create steric stabilization, in which adsorbed polymer chains on adjacent particles intermingle thereby hindering the attraction between the particles. Electrostatic stabilization can also occur, which means charge created on the particles will cause an electrostatic repulsion, thereby aiding in the stabilization of the suspension. It was found that a possible effect of using a polymeric dispersant in an oxide slurry is the lowering of the zeta potential.[1]

Zeta potential can be useful in prediction the behavior of particles in a multi-component paste. The isoelectric point of a material can be determined by measuring the zeta potential as the pH of the suspension is varied. Comparison of the isoelectric points of two materials with different surface charge behavior can indicate whether the materials will have an electrostatic attraction to one another multi-component system.

Fatty acid polymers are composed of a hydrophilic carboxyl group and a hydrophobic carbon chain. These fatty acids tend to have a low solubility in water, thus complicating their use in manufacturing formulations which utilize water as the liquid medium. The saponification process exchanges the hydrogen in the carboxyl group for a sodium ion. This changes the nature of the bonding (from polar OH to ionic O-Na) and switches the solubility properties of the acid from being insoluble in water, soluble in methanol, before the saponification to the reverse afterwards.

The conversion of a fatty acid into a sodium salt can be achieved by the alkali refining of the acid using sodium hydroxide in a methanol solution.[2] This method has been used to create soap stock from a replicated soybean oil to be further converted into a fatty acid methyl ester.

Hydrocarbon chain length has been found to have an effect on the amount of polymeric adsorption to inorganic particles.[3] A study analyzing the adsorption of decyltrimethylammoniums of different chain lengths onto clinoptilolite found that greater adsorption occurred for hydrocarbons with longer chain lengths. This study used adsorption isotherms and zeta potential to explore ion exchange and hydrophobic interactions at the clinoptilolite/water interface.

The effect of combining fatty acids of different chain lengths has been explored. The fatty acids used were a saponified stearate and caprylate. These acids were included as lubricants in a paste. Torque and capillary rheometry both show a lowering of the yield stress when the acids were combined as compared to the same quantity, by mass, of either acid individually.[4]

UV spectroscopy has been found to be an effective method of determining the quantity of some polymers in solution. There are two ways of using this technique depending upon the substrate the polymer is intended for. Poly (o-methoxyaniline) was applied to a glass substrate. UV absorption by the polymer was found to correlate to the quantity of polymer occupying the surface of the glass.[5] In the case of particle coating, the method of determination is subtractive. Hydroxypropyl methyl cellulose was adsorbed to various particle surfaces. Various amounts of the polymer were used and the supernatant of each solution analyzed to create adsorption isotherms for the polymer with each inorganic.[6]

Torque rheometry is useful in examining the rheological properties of a paste. This technique focuses on the transition from dry ingredients and water, to a single paste system. On a graph resulting from torque rheometry, this transition is examined by using the peak torque and the time to peak. The peak torque is the maximum torque required to plasticize the paste. The time to peak torque indicates how long it takes for this transition to occur. After peaking, the torque relaxes to a stable value. Torque values are related to viscosity of the paste.

Dynamic stress rheometry can be performed on either a slurry, using a vane geometry, or an extruded paste, using a parallel plate geometry. DSR involves the set oscillation of the equipment geometry while in contact with the sample. The resultant stress is then measured yielding two parameters, G' and G''. These parameters can then be used to determine the required oscillation to cause the sample to yield. This yield point marks the transition of the sample from elastic to viscous flow. This transition can be abrupt or drawn out depending on the nature of the bonding of the agglomerates in the sample.

EXPERIMENTAL PROCEDURE

Saponification

Sodium based soap is synthesized by adding a fatty acid (ex. stearic acid) to sodium hydroxide in a solution of methanol. These fatty acids have a 1:1 chemical reaction with the sodium hydroxide. The general reaction is

$$NaOH + FA \rightarrow NaFA + H_2O$$

Sodium hydroxide is added to methanol and agitated at an increased temperature. Evaporated methanol is recycled via a reflux condenser. When the sodium hydroxide is completely dissolved, the fatty acid is added in proportion to the ratio of the molecular weight of the fatty acid over the molecular weight of the sodium hydroxide. Agitation of this mixture then continues until the reaction has completed and the fatty acid salt has precipitated out. The remaining methanol is then allowed to evaporate off, leaving the salt. Table I shows a range of fatty acids, their carbon chain lengths and molecular weights.

Thermal Analyses

Differential thermal analysis was performed on the saponified fatty acids, sodium caprylate, sodium laurate, sodium palmitate, sodium stearate, and sodium oleate using the Perkin-Elmer TGA7. The samples were analyzed as the temperature was ramped from room temp to about 950°C at a rate of 10°C/minute. Thermal gravimetric analysis was also performed on sodium caprylate, sodium laurate, sodium stearate, and sodium oleate using the Perkin-Elmer DTA7. These samples were examined as they were heated from room temperature (25°C) to about 800°C. A ramp setting of 5°C/minute was used during the sample analysis. Both of these analyses were performed in an air atmosphere.

Torque Rheometry

Measurements were taken on a Haake Rheometry with a dual sigma blade configuration at 50 rpm for 15 minutes. Nine batches were analyzed through torque rheometry. Five of these batches included sodium stearate, sodium oleate, sodium palmitate, sodium laurate, and sodium caprylate. The remaining four batches each contained sodium stearate combined with one of the remaining fatty acids. All of the batches contained A-2 alumina, 3% weight methyl cellulose binder, 2% one or both of the saponified fatty acids, and 20% water. A-2 alumina was used in the rheological testing because its particle size distribution was the most convenient for analysis in the equipment utilized.

Zeta Potential

Zeta potential measurements were taken with a Dispersion Technology, DT-1200 electro-acoustic spectrometer. Samples consisted of 150 ml of slurry with 10% inorganic by volume. The DT-1200 was first used to analyze the particle size distribution in the slurry. This information was then used along with the measured colloid vibration current in the calculation of the zeta potential. The pH of the slurry was varied by titrating in known quantities of 1 M HCl or 1 M KOH. The inorganics analyzed in this test were alumina and talc. Each was tested individually.

Dynamic Stress Rheometry

Dynamic stress rheometry was performed on a TA Instruments, AR-1000-N Rheometer. Sample prep included dissolving fatty acid in water then adding it to the inorganic components. This mixture was then compounded in a Brabender Plasticorder for 20 minutes before being extruded as 1.5 cm rods from a Loomis extruder. A 0.5 cm cylinder was then cut from the center of the rod to be tested on the rheometer.

A parallel plate geometry was used on the TA rheometer. This geometry consists of a striated base plate on which the sample is centered. Another striated plate, parallel to the base, is then lowered until it is making contact with the sample. For this experiment, the range of oscillations the sample underwent was between 100 and 100,000 Pa. The samples analyzed with this technique were each composed of a 50/50 mixture of alumina and talc along with 3% methyl cellulose binder, 20% water, and 0, 0.5%, or 1% fatty acid. The fatty acids analyzed with this technique were sodium stearate and sodium laurate.

RESULTS & DISCUSSION

Thermal Analyses

In the thermal gravimetric analysis of fatty acids, shown in figure 1, the weight loss for saturated fatty acids appears very similar—a large, smooth weight loss over a 60-70 °C range, followed by a slower loss of the remaining 2-9% of the weight over the next 50-200°C. The shape of these curves is similar for all saturated fatty acids, but the temperature at which weight loss occurs increases with molecular weight, for most part.

Differential thermal analysis of these fatty acids show two main events. There is an endothermic peak at the melting point. However, there is no second endothermic peak at the boiling point as might be expected. Instead, there is a small exotherm that corresponds to the main weight loss of the fatty acid. The weight loss and exotherm occur at temperatures below the boiling points of the fatty acids. This indicates that the fatty acid is oxidizing ("burning off") before it reaches its boiling point. The boiling point of all fatty acids is well above the temperature at which 50% of the total weight loss has already occurred. For the saturated fatty acids, more than 95% of the total weight loss occurs before the boiling point. From this, it can be concluded that saturated fatty acids do not evaporate in air, but rather oxidize.

In the thermal analysis of oleic acid, an unsaturated fatty acid, the initial weight loss accounts for about 80% of the total weight loss. The slope of this weight loss is slightly less than that of the saturated fatty acids, as well. The second weight loss, of the remaining 20 weight percent, occurs over a temperature range of more than 200 °C (285-490 °C). Since the DTA shows an exothermic peak over this weight loss, it can be concluded that this also is a result of oxidation. The difference in slope between saturated and unsaturated fatty acids indicates that the mechanism for oxidation must be slightly different.

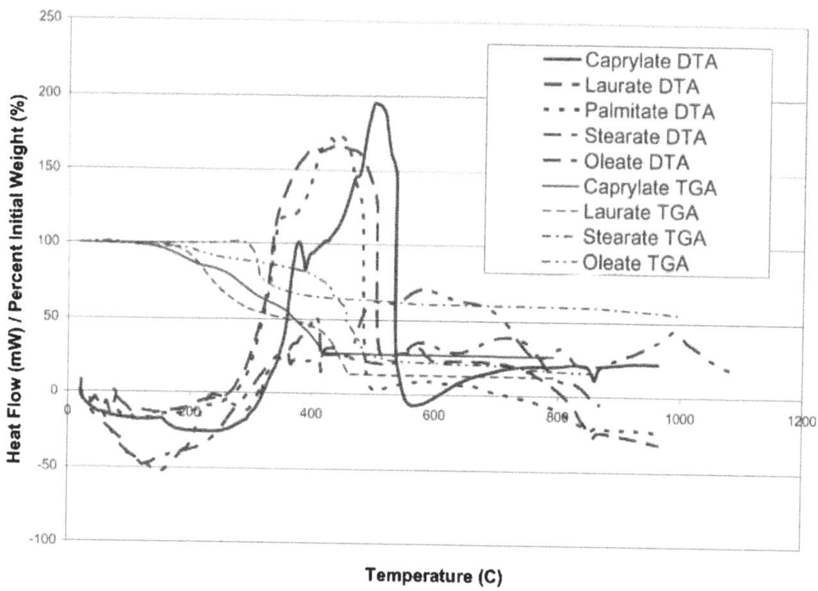

Figure 1 Thermal Analysis of Saponified Fatty Acids

Torque Rheometry

Figures 2 and 3 show torque rheometry results for single and dual organic compositions, respectively. Figure 2 shows that the fatty acids with longer chains, sodium stearate and sodium oleate (both 18 carbons), have lower peak torque values than the shorter chain acids, sodium caprylate (10 carbons) and sodium palmitate (16 carbons). The exception to this trend is sodium laurate (12 carbons), which has the lowest peak torque value.

In figure 3, the sodium stearate/sodium palmitate paste has the highest peak torque value. The sodium stearate/sodium caprylate paste has a similar curve to the sodium stearate/sodium laurate paste. The lowest peak torque value belongs to the sodium stearate/sodium oleate combination. Comparing figures 2 and 3 shows us that, with the exception of the sodium stearate/sodium palmitate paste, all of the pastes with sodium stearate and another fatty acid have lower peak torque values than sodium stearate alone. The peak torque values for each paste can be seen in Table I.

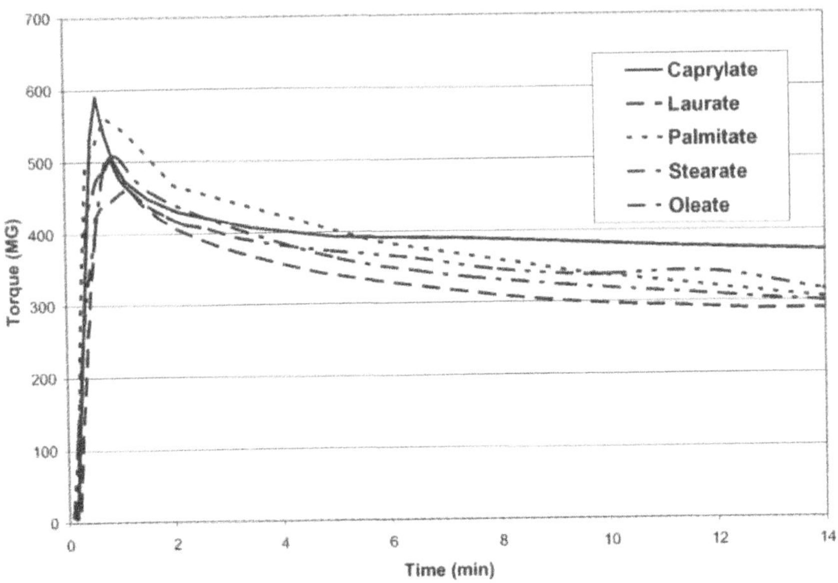

Figure 2 Torque Rheometry of Fatty Acid in an Alumina Paste

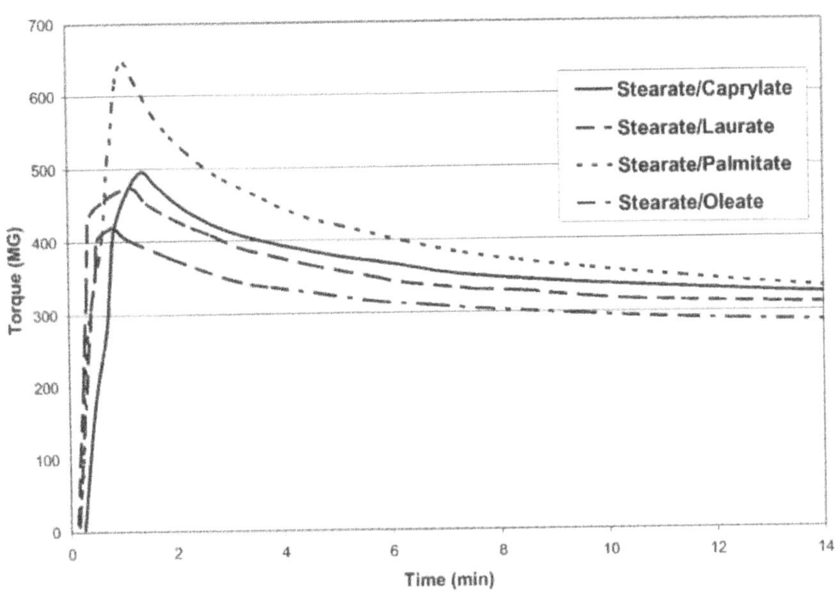

Figure 3 Torque Rheometry of Combined Fatty Acids in an Alumina Paste

Table I Peak Torque values for Fatty Acids in an Alumina Paste

Fatty Acid Salt	# of Carbons in Chain	Peak Torque (MG)
Sodium Caprylate	10	591
Sodium Laurate	12	456
Sodium Palmitate	16	562
Sodium Stearate	18	507
Sodium Oleate	18 (double bond)	501
Sodium Stearate/Sodium Caprylate	18/10	496
Sodium Stearate/Sodium Laurate	18/12	474
Sodium Stearate/Sodium Palmitate	18/16	648
Sodium Stearate/Sodium Oleate	18/18	419

Zeta Potential

Figure 4 shows the results of observing the effect of pH on the zeta potential of alumina and talc inorganics. The isoelectric point of alumina was found to occur at a pH of approximately 8.6. The isoelectric point of talc was found to be much lower, around a pH of 4.8. The significance of knowing the isoelectric points for these two inorganics is to determine whether electrostatic attraction will occur when they are in a slurry together. The pH of a paste created by using 50% of each of these inorganics is 9. This indicates that there will not be a tendency to have a strong charge on the alumina particles in the paste.

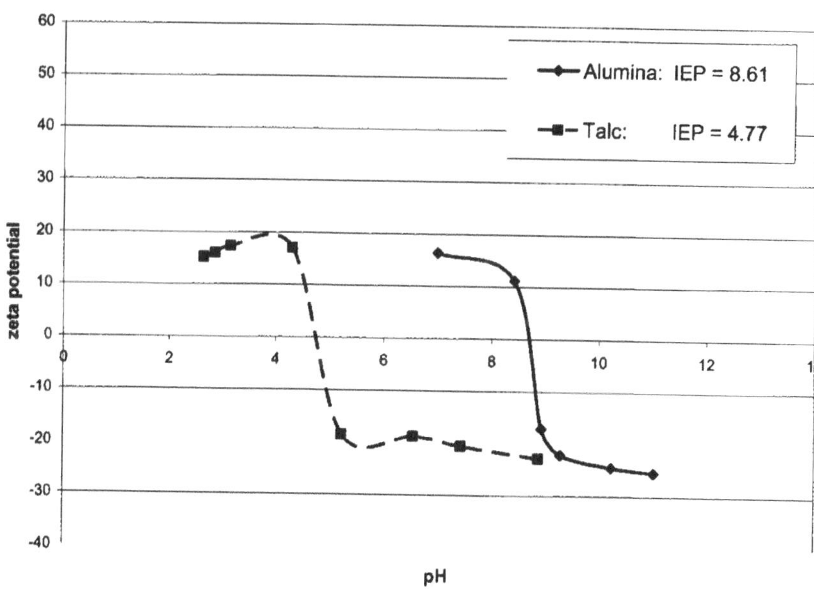

Figure 4 Isoelectric Point Determination for Alumina and Talc

DSR

Figures 5 and 6 show the G' curves for sodium stearate and sodium laurate, respectively, as determined by dynamic stress rheometry. Some of these G' curves are cut off before a distinctive yield point could be determined. This is due to limitations in the amount of oscillatory stress which the sample could withstand before becoming dislodged from the base plate and compromising the test results. Even without distinctive yield points, this data is still valuable in highlighting trends. The samples without any additive show an inclination to have the highest yield points. The G' values then begin to drop with the addition of fatty acid salt. For the values of fatty acid used in the pastes, the trend shows that greater additions of fatty acid lead to lower yield stresses. This trend is consistent for both sodium stearate and sodium laurate additions. Comparing figures 5 and 6 show that the additions of sodium laurate are more effective at lowering the yield stress of the paste than the sodium stearate additions.

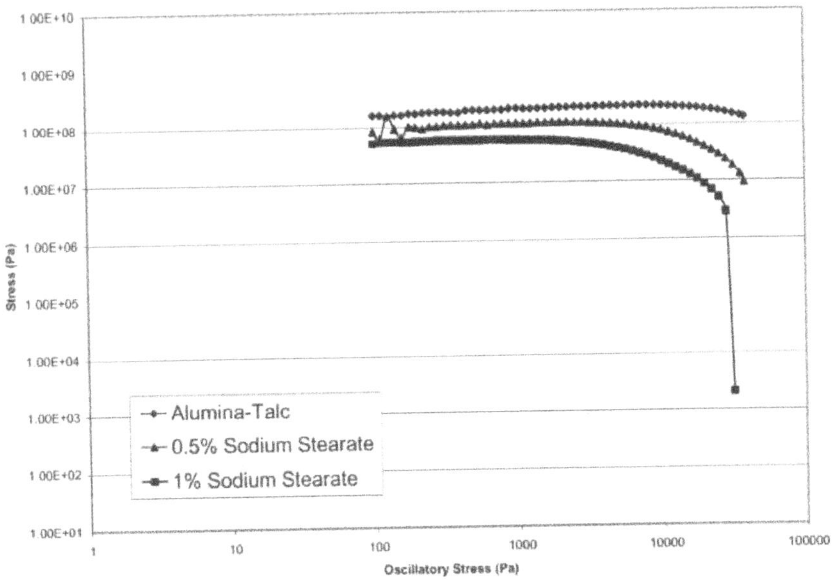

Figure 5 Dynamic Stress Rheometry of a paste with Sodium Stearate Additions

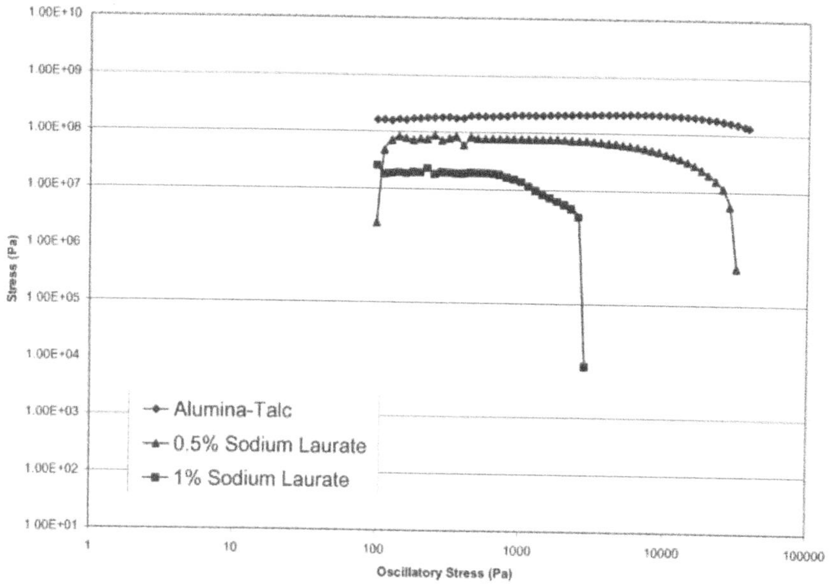

Figure 6 Dynamic Stress Rheometry of a Paste with Sodium Laurate Additions

CONCLUSION

Effects of hydrocarbon chain length on the thermal behavior of a saponified fatty acid have been shown with differential thermal analysis and thermal gravimetric analysis. Torque and dynamic stress rheometries have been used to highlight the effects of mixing two fatty acid salts of different chain lengths on the rheology of a single component paste as well as the effects of a single fatty acid on the rheology of a multi-component paste. Dynamic stress rheometry has shown that greater quantities of lubricant can lower the yield stress of a multi-component paste. The torque rheometry results indicate that mixing two fatty acid salts with different chain lengths can lead to lower peak torque values than a single fatty acid salt. Determination of the isoelectric points of alumina and talc lead to the conclusion that there will not be strong electrostatic attraction between the particles.

REFERENCES

[1]D.W. Fuerstenau, R. Herrera-Urbina, and J.S. Hanson, "Adsorption of Processing Additives and the Dispersion of Ceramic Powders," *Ceramic Powder Science*, **2**, 331-351 (1988).

[2]M.J. Haas, S. Bloomer, and K. Scott, "Simple, High-Efficiency Synthesis of Fatty Acid Methyl Esters from Soapstock," *Journal of the American Ceramic Society*, **77**, no 4, 373-9 (2000).

[3]B. Ersoy and M.S. Celik, "Effect of Hydrocarbon Chain Length on Adsorption of Cationic Surfactants onto Clinoptilolite," *Clays and Clay Minerals*, **51**, no 2, 172-81 (2003).

[4]C.R. August, R.A. Haber, L.E. Reynolds, "Comparison of Sodium Caprylate and Sodium Stearate and the effects of their Hydrocarbon Chain Length on Adsorption Behavior and Alumina Paste Rheology," *Ceramic Engineering Science Proceedings*, (2006)

[5]M. Raposo, L.H.C. Mattoso, O.N. Oliveira Jr., "Adsorption of poly(0-methoxyaniline)," *Thin Solid Films*, 329, 739-42 (1998).

[6]C.B. Sawyer, J.S. Reed, J.A. Lewis, "Adsorption of Hydroxypropyl Methyl Cellulose in an Aqueous System Containing Multicomponent Oxide Particles," *Journal of the American Ceramic Society*, 84, no 6, 1241-29 (2001)

Eco-Materials Processing and Design
Edited by Koji Watari and Soo-Wohn Lee

GELCASTING PROCESS WITH BIOPOLYMER IN NATURAL OIL FOR MAKING CERAMIC SPHERE

Thammarat Panyathanmaporn*, Rung-Arun Chumnanklang, Sitthisuntorn Supothina, Angkhana Jaroenworaluck
National Metal and Materials Technology Center, 114 Thailand Science Park, Paholyothin Rd., Klong 1, Klong Luang, Pathumthani 12120 Thailand

ABSTRACT

Gelcasting is a forming method, which is based on the consolidation of fluid well-dispersed slurry into a rigid body without liquid removal. The gel casting process consists of the dispersion of ceramic powders in the monomer solution to form a fluid that is subsequently cast into the non-porous mould. At the elevating temperature or by acting by catalyst, initiator and a cross-linking agent, the monomers polymerise to form a strong, cross-linked polymer-solvent gel and permanently immobilised the ceramic slurry in the desired shape. However, commercial monomers, crosslinkers and initiators are toxic. In this work we are interested in using biopolymers, *i.e.* starch, as binders because it is environmentally friendly, easy to burn out and very economic. To make ceramic spheres, Al_2O_3 slurry mixed with starch solution to form a fluid is prepared. By dropping small volumes of Al_2O_3-starch solution in a vegetable oil at the high temperature, the starch forms a strong, cross-linked gel and immobilises the Al_2O_3 powders in the spherical shape. Al_2O_3 spheres were dried and then sintered at 1700 °C for 1 hour. The sintered porous Al_2O_3 spheres exhibit a broad diameter range of 50 to 920 μm with a mean sphere size of 325 μm. The measured porosity area of Al_2O_3 spheres is 48%. SEM examination reveals that the surface appearance shows a wrinkled structure integrated with coarse pores. Also, clay spheres are shown.

INTRODUCTION

Gel casting is an attractive technique for molding ceramic powder based on a synthesis derived from traditional ceramics and polymer chemistry.[1 2 3] Gel casting is a near-net shape forming process similar to slip casting and injection molding. The process consists of the dispersion of ceramic powders in the monomer solution to form a fluid that is subsequently cast into the non-porous mould. At the temperature of 60 °C or by acting by catalyst, initiator and a cross-linking agent, the monomers polymerise to form a strong, cross-linked polymer-solvent gel and permanently immobilised the ceramic slurry in the desired shape. The advantages of gel casting include capability of producing complex parts like injection molding. It is ease of implementation due to its similarity with well-established processes like slip casting (low capital equipment cost). Also, this technique is possibility to use low-cost mold material and low organic contents meaning ease of binder burnout. Moreover, gel casting ceramic shrinks uniformly. Thus, the mold can be designed to compensate for shrinkage so that a ceramic part of the desired shape and size can be obtained. Another advantage of the gel casting is the process called 'green-machining', which is used when the machining is desired to improve on the final shape. After the 'green body' stage, a gelcast part is molded and dried. It is strong and soft enough to be machined quickly by less costly carbon steel tools. Moreover, the final product shows very homogeneous material properties. Finally, this generic method can apply for both ceramic and metal powders.

However, gelcasting via polymerization of monomer is environmental and health threats since monomers generally used are toxic. Alternatively, there are many biopolymers currently developed for gelcasting process to replace toxic substances and forming gel under suitable condition, such as starch and gelatin. Besides being a health and environmental friendly, biopolymer gels are highly hydrated (solvated) and form homogeneous three-dimensional network across the gel body. Therefore, the biopolymer is one choice to use as binders for ceramic processing.

Starch is a natural biopolymer employed as binder for gel-casting process.[4] Starch consists of two polysaccharide type, amylase (linear) and amylopectin (branches). When starch is heated in water, the starch granules swell and lose their semi crystallinity. This phenomenon is generally referred to as gelatinization.[5] This gelatinization causes a network structure and leads ceramic powders to stick together. Due to its gelatinization phenomenon, the starch is acted as binder. After binder burn-out and sintering process, porous ceramic structure is obtained. Its porosity is related to amount of added starch.[6]

With favorable gelatinization of starch, gelcasting process is beneficial to form ceramic in many shapes but spherical shape hasn't been fabricated by this process. Several processes have been reviewed for preparing spherical shapes of ceramic such as rotating pan granulation, spray-drying, liquid phase molding, oil molding and hydrocarbon ammonia molding [7]. Hence in this work, interested in alumina spheres forming by aqueous gelcasting technique in hot oil is focused. The microstructure and dimensions of obtained alumina spheres were revealed.

EXPERIMENTAL AND RAW MATERIALS

The materials and techniques used for preparation of the kaolin spheres in this work are described. The raw materials were characterized by Thermal analysis and Scanning Electron Microscopy (SEM) whereas alumina spheres are studied by Scanning Electron Microscopy. Alumina used in this study was purchased from Compound Clay Co., Ltd., Thailand. Cassava starch with 13 % of moisture was supplied by Thai Wah Food Products Public Co., Ltd,. Refined palm oil was supplied by Emerald Brand Cooking Oil Public Co., Ltd. All raw materials were used as received.

Thermal analysis

Differential scanning calorimetry in this study was Perkin Elmer, DSC7. The ingredient of sample, 0.5g of starch dissolved into 10 ml of water, was placed in aluminum pan. The analyses were performed under nitrogen flow (60 ml min^{-1}) at a heating rate of 10°C min^{-1}. Starch powder as binder were analyzed by Thermogravimetric analysis (Perkin Elmer, TGA7), at a rate of 20°C min^{-1} from ambient temperature to 800°C. The analyses were performed under air atmosphere.

SEM observation

Microstructure of the green and sintered alumina spheres were studies by a SEM (JSM-6301F, JEOL, Japan). Samples were dried at 110°C for 24 hours and then coated with gold for

100 seconds after cooling in desiccator. The porosity of the sintered alumina spheres and the size of 200 sintered spheres were measured by Image Analysis Program (Omnimet 1.4 beta).

Slip preparation

Typical batch composition is consisted of 100 g of alumina and 2 ml of dispersant, Darvan C, which is added into 30 ml of distilled water. The alumina slip was milled by ball milling overnight and then mixed with 7 g of cassava starch equaling to 7%wt of alumina.

Spherical forming

The obtained homogeneous slip was added into vegetable oil to form droplets with emulsion state, and thereafter, slow stirring of these droplets resulted in fine droplets. These fine droplets were carefully took up, and dropped into hot oil at about 140-180 °C with soaking time of 15 min. Then the green spheres were separated from oil and placed on the tissue paper to remove the excess oil before drying at 100°C for 2 hours.

Burnout and sintering

Green spheres were conducted in a furnace for sintering. Two continuous steps for sintering process were applied. They were heated up to 600°C at a slow rate (1°C min^{-1}) with 1 hour holding, and then were further heated up to 1700°C at faster rate (5°C min^{-1}) followed by holding for 1 hour at this temperature.

RESULTS AND DISCUSSION

Gelatinization of starch and its burn-out behavior

The gelatinization of starch was studied by Differential scanning calorimeter.[5] Thermogram in Figure.1 shows relationship between heat flow and temperature. The thermogram presents the onset and peak temperature at 66.17 and 69.52 °C, respectively. The temperature range between the onset and peak temperature indicates the gelatinization temperature range of starch.[8] This gelatinization temperature range is very similar to the gelatinization temperature measured by the Rapid visco analyzer.[9] Moreover, Figure 2 shows the microstructure of the gelled starch. The starch granules in the aqueous solution heated up to gelatinization temperature (70°C) can intact with neighboring molecules to form the network of gel as shown in Figure 2 (a) and (b. This occurrence may be explained by the work done by Gallent *et.al.*[10] During gelatinization, amylose leached out of the granule formed a network structure by filling the inter-granular space. However, in spherical forming step, the temperature of hot oil was set in the range of 140-180 °C in order to ensure that starch in ceramic slip can be gelled completely while emulsion droplets are heated in hot oil.

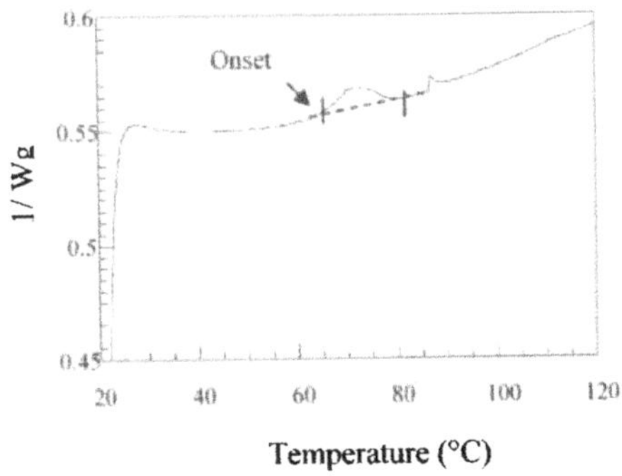

Figure 1 DSC pattern of gelatinization of cassava starch

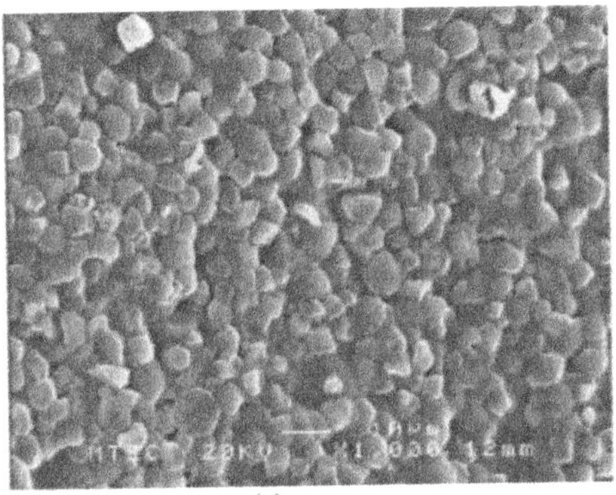

(a)

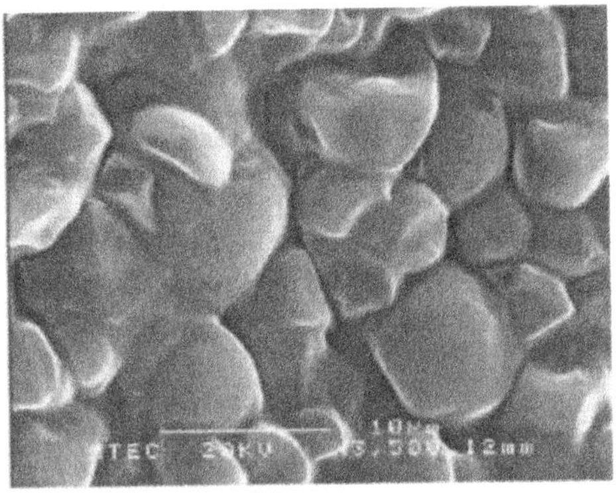

(b)

Figure 2. SEM micrographs of cassava starch gel heated at 70°C taken at 1,000 (a) and 3,500 (b) of magnification

The weight loss of starch during binder burn-out step was studied by Thermogravimetric analysis. The TGA curve in Figure 3 shows that the water was removed in the range of 50-180 °C whereas starch containing the linear and the branch exhibits two different decomposition at 280 and 405 °C, respectively which are similar to the study done by Lyckfeldt and Ferreira.[11] Also all the organic component can be burn out completely at 550 °C. The ease of burn-out starch and sintering step is favorable to ceramic forming as starch, which is used as a binder, doesn't leave any trace.

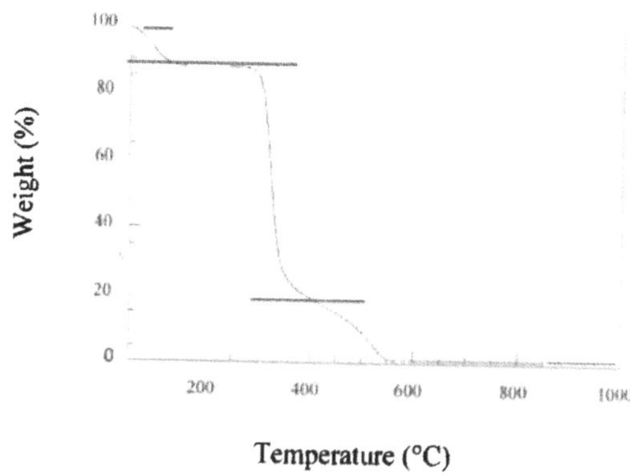

Figure 3 TGA of cassava starch

Microstructure of green spheres

Green alumina spheres were fabricated by gelcasting using starch as binder in hot oil. The microstructures of the obtained green spheres are illustrated in Figure 4 with 250 times of magnification. SEM images show intact spheres prepared successfully with this technique. A few broken spheres were observed. The spheres consisted of closely packed aggregates of alumina particles with some trace on their surface. Pores up to 10 µm were observed on the surface.

Dimension and microstructure and of sintered spheres

An image of sintered alumina spheres is present in Figure 5. Most of the spheres remained intact after sintering at 1700 °C. The sintered spheres did not collapse even though all the starch had been completely burnt off. The dimension of sintered spheres was analyzed by Image Analysis Program (Omnimet 1.4 beta). The particle size distribution of sintered spheres is shown in Figure 6. The sintered spheres had a broad range of diameter size of 50 to 920 µm with a mean size of 325 µm. Also, Microstructure of the sintered spheres is shown in Figure 7. After sintering, most population of the sphere exhibited a surface appearance consisting of a wrinkled alumina structure integrated with the coarse pores. The internal structure of the sphere shows a similar structure to sphere surface as shown in Figure 8. This occurrence can be explained as follow. During the binder burnout step, the gelled starch, which connected alumina particles together, was burnt into the gas from the green sphere and hence left the pore between particles. The porosity was measured with the same program of dimension. Figures 9 (j) and (k) show the analysis of porosity with calculation of different color of interesting area. The bright object is the area of pores whereas the dark object is the area of alumina. The average porosity obtained is 48%.

The success of this alumina sphere forming method led to the fabrication other materials such as clay. Kaolin clay was then used to form a kaolin sphere following the methos employed for making alumina sphere. Figure 10 shows the image of the sintered kaolin spheres. The intact kaolin spheres are similar to the alumina spheres.

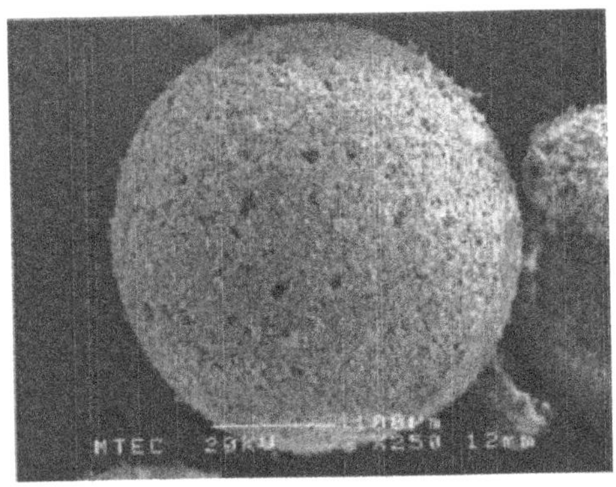

Figure 4 SEM micrograph of green alumina spheres

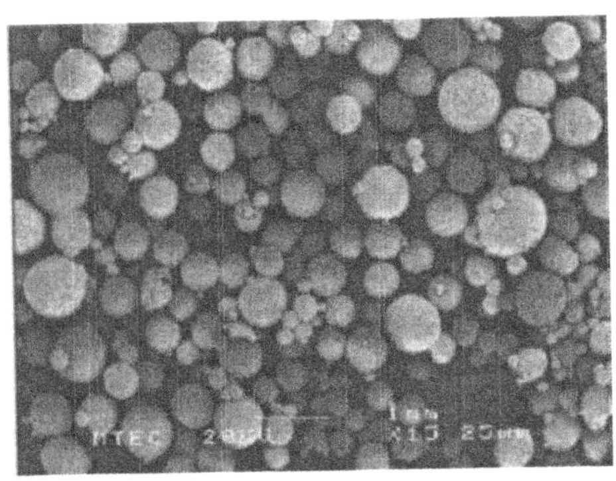

Figure5 SEM micrograph of alumina spheres after sintering at 1700 °C

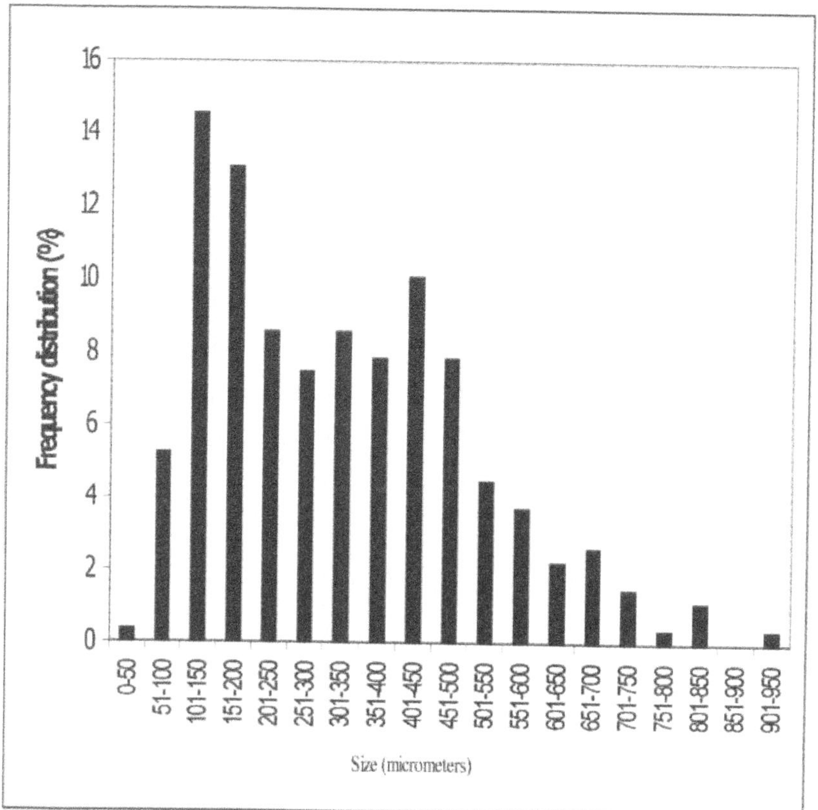

Figure 6 Particle size distribution of the sintered alumina spheres

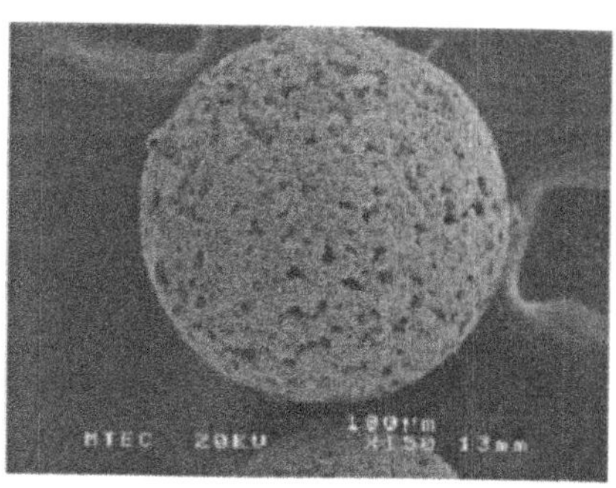

Figure 7 SEM micrograph of the sintered alumina spheres

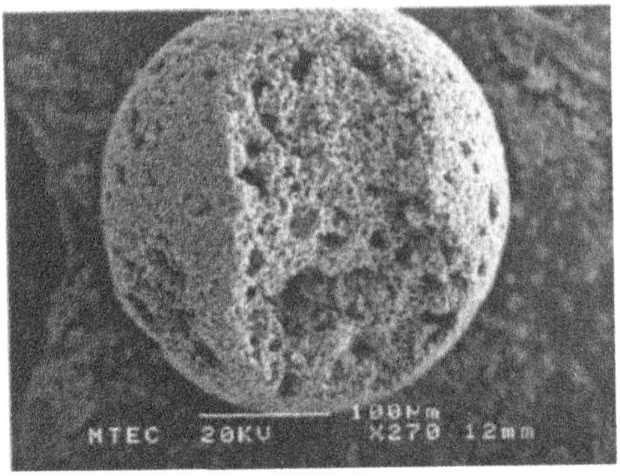

Figure 8 SEM micrograph of internal structure of the sintered alumina spheres

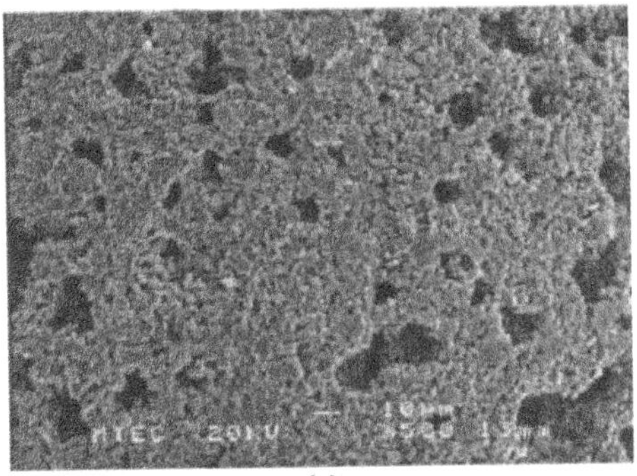

(a)

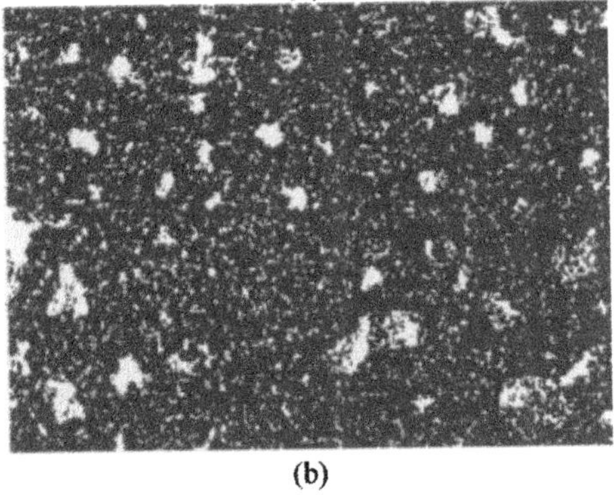

(b)

Figure 9 SEM micrographs of the sintered alumina spheres (a) Micrograph of sphere surface before porosity analysis (b) Micrographs of image analysis for porosity.

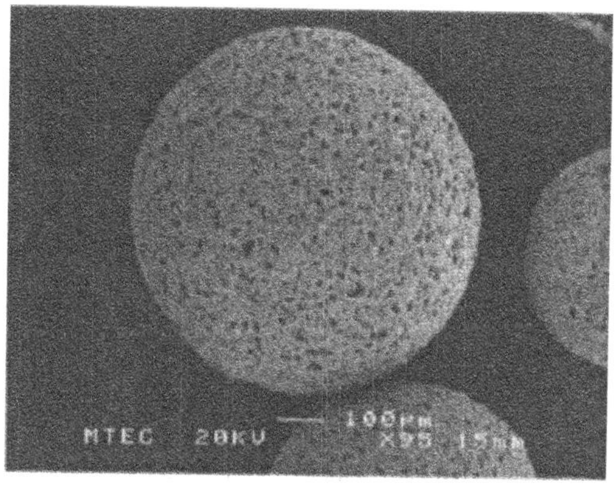

Figure 10 SEM micrograph of the kaolin spheres after sintering at 1300 °C

CONCLUSIONS

The preparative method of alumina and kaolin spheres has been successfully developed by using gelcasting technique. An environmental friendly, inexpensive starch was used as a binder in place of a more toxic, expensive monomer. The green spheres were sufficiently rigid and could be handled. The favorable properties of starch in terms of gelling ability, low cost and low toxic making aqueous gel casting with biopolymer in hot oil are an alternative method to produce the spherical ceramics. The average porosity of the alumina sphere is 48%, while the dimension of these spheres varies between 50 to 920 μm with mean diameter of 325 μm.

ACKNOWLEDGEMENT

This work has been financially supported by the National Metal and Materials Science Center (MTEC), funding number MT-B-47-CER-07-169-I. Also, this work is one part of the patent pending number TH0974341. The authors also wish to thank MTEC's technicians who conducted all the characterization.

REFERENCES

[1]OO. Omatete, MA. Janney and SD. Nunn, "Gelcasting: From Laboratory Development toward Industrial Production" *J. Eur. Ceram. Soc.*, **17**, 407-413 (1997).

[2]R. Gilissen, JP. Erauw, A. Smolders, E. Vanswijgenhoven and J. Luyten, "Gelcasting a Near Net Shape Technique" Materials and Design., **21**, 251-257 (2000).

[3]WM. Sigmund, NE. Bell and L. Berström, "Novel Powder Processing Methods for Advanced Ceramics" *J.Am.Ceram.Soc.*, **83**(7), 1557-1574 (2000).

[4]R Chumnanklang and T Panyathanmaporn. "Effect of bio-polymer on microstructure of gelcast alumina" *Proceeding of The 4th Asean Microscopy Conference and the 3rd Vietnam Conference on Electron Microscopy*, Vietnam National University: Hanoi, Vietnam, 181-182 (2004).

[5]Ch Chauncey, David LK, Edwin LTh. *Biodegradable Polymers and Packaging*, U.S.A., 65-67 (1993).

[6]T Panyathanmaporn, A Fuongfuchat, T Leejarkpai, P Surunchanairasakul and Chumnanklang R. "Gelcasting of porous alumina by using cassava starch as binders" *Proceeding of The 29th Congress on Science and Technology of Thailand*, Khon Kean University, Thailand, 213 (2003).

[7]Z.R Ismagilov, R.A Shkrabina, and N.A Koryabkina, " New Technology for Production of Spherical Alumina Supports for Fluidized Bed Combustion". *Catalyst Today.*, **47**; 51-71 (1999).

[8]JJM Swinkels,. *Starch Conversion Technology*, New York, 15-45 (2003).

[9]T. Panyathanmaporn, A. Fuongfuchat, T. Leejarkpai and P. Surunchanairasakul In "Preliminary Study of Use Biopolymer as Binders in Ceramic Gel Casting", *Proceeding of The 2nd Thailand Material Science and Technology: Materials Science and Technology for a Sustainable Development of Thailand*, Kasetsart University, Thailand, 255 (2002).

[10]DJ. Gallant, B. Bouchet and PM. Baldwin, "Microscopy of Starch Evident of a New Level of Granule Organization" *Carbohydrate Polymers.*, **32**, 177-191 (1997).

[11]O Lyckfeldt and JMF Ferreira, "Processing of Porous Ceramics by Starch Consolidation" *J. Eur. Ceram. Soc.*, **18**; 131-140 (1998).

Eco-Materials Processing and Design
Edited by Koji Watari and Soo-Wohn Lee

LOW-TEMPERATURE SINTERED LZSA CELLULAR GLASS CERAMICS

E. Sousa[1], C.R. Rambo[1], D. Hotza[1*], A.P. Novaes de Oliveira[1]

[1]Graduate Program on Materials Science and Engineering – PGMAT
Department of Chemical Engineering – EQA
Federal University of Santa Catarina
P.O. Box 476, 88040-900 Florianópolis, SC, Brazil

T. Fey[2]
[2]Department of Materials Science, Glass and Ceramics
University of Erlangen-Nuremberg
Martensstrasse 5, D-91058 Erlangen, Germany

ABSTRACT

Commercial polyurethane foams with a monomodal pore size distribution were used to produce LZSA glass ceramic foams by the polymeric sponge method. A suspension containing LZSA glass ceramic, bentonite and sodium silicate, was prepared in water and isopropanol media to impregnate the polymeric foams by dip coating. The suspension was characterized by rheological measurements. The effect of the solvent on the microstructure and physical properties on the LZSA foams was also evaluated. The cellular microstructure of the glass-ceramic foams was characterized by scanning electron microscopy (SEM) and micro-computer X-ray tomography (μ-CT). The LZSA foam prepared with isopropanol suspension exhibited higher mechanical strength under compression than those prepared with water.

INTRODUCTION

In the recent years the interest in materials with cellular structure, such as foams, reticulated and biomorphic ceramics has increased due to their specific properties such as low density, low thermal conductivity, thermal stability, high surface area and high permeability [1-3]. These properties make ceramic foams suitable for a wide range of technological applications, for example, catalyst supports, filters for molten-metals and hot gases, thermal insulators, refractory linings, and biomaterials [4,5].

Ceramics foams are a special class of porous materials comprised of large voids (cells), with linear dimensions approximately ranging from a few micrometers to few millimeters (10 μm – 5 mm) and possessing a geometry similar to that of a tetrakaidecahedron. Foams are usually sub-divided into two categories, open cells and closed cells [6,7].

Various processing routes have been proposed for production of ceramic foams, including polymeric sponge [8,9], foaming agents [10,11] or space holder method [12]. The fabrication method determines the range of porosity, the pore size distribution and the pore morphology.

The polymeric sponge method or replication method was patented in 1963 by Schwartzwalder and Somers [8]. It is a simple, inexpensive and versatile way for producing ceramic foams. This method consists of the impregnation of a polymeric sponge with ceramic slurries containing ceramic particles and appropriate binders followed by a heat-treatment which leads to the burning out of the organic body (sponge) and to the sintering of the ceramic skeleton.

Usually the ceramic foams made by polymeric sponge method show high porosity (70-95%) and high permeability. Among a wide range of ceramics that can be made by this method, zirconia, silicon carbide and alumina are the most common [8-10]. However, only few works dealt with the production of cellular glass-ceramics [13]. Glass ceramics belonging to the Li_2O-ZrO_2-SiO_2-Al_2O_3 (LZSA) system exhibit good chemical, thermal shock resistances, and extraordinary low thermal expansion coefficient [14], which make them suitable for applications where no shrinkage during thermal treatment is required.

The present work deals with the rheology of the glass ceramic precursors and the influence of the rheological parameters on the synthesis of glass ceramics LZSA foams prepared by the replication method.

EXPERIMENTAL PROCEDURE

The LZSA glass ceramics were prepared from Li_2CO_3, $ZrSiO_4$ and SiO_2 and spodumene as raw materials. Details of the preparation and characterization are described elsewhere [15]. The physical and chemical characteristics of the powders are shown in Table 1.

Table 1. Chemical and physical properties of the LZSA glass and bentonite powders.

	Specific surface area (m^2/g)	**Mean size distribution (d_{50}, μm)**	**Density (g/cm^3)**	**Main crystalline phases**
LZSA	4.95 ± 0.05	3.22 ± 0.05	2.63 ± 0.05	Amorphous and zircon
Bentonite	29.45 ± 0.05	4.47 ± 0.05	2.49 ± 0.05	Montimorilonite and quartz

The slurry was prepared with two different solvents, water and isopropanol, containing 60 vol.% parent glass powder (LZSA), 1-10 vol.% bentonite (Colorminas, Brazil) used as a binder, and 1 vol.% sodium silicate (Merck, Natronwasserglas 105621, 7.5-8.5% Na_2O, 25.5-28.5% SiO_2), used as dispersant. The solvent, water or isopropanol, was first mixed with sodium silicate in a plastic bottle with alumina ball as grinding media for 12 h. Subsequently, the glass powder was added to the slurry and then milled for another 12 h. Bentonite was added to the resulting slurry and kept milling for another 12 h.

Commercial polyurethane foams with a monomodal pore size distribution (55 ± 5 pores per inch) and porosity of 95 ± 1% were used as templates. The polymeric foams were cut in pieces of approximately 2.5 x 3.0 x 2.0 mm and immersed in the LZSA parent glass ceramic slurry. The impregnated foams were slightly compressed to remove the excess slurry and dried at room temperature for 24 h. After drying, the samples were subjected to the heat treatment, which was performed in an electrical furnace (EDG 1800) in air atmosphere. The samples were first heated at 1°C min^{-1} to 400 °C to burn out the polymeric foam. Subsequently, the samples were heated to 700°C at 5°C min^{-1} to promote the sintering of the glass powder and finally were heated at 900 °C with the same heating rate for crystallization. The system was inertially cooled.

The rheological behavior of slurries was analyzed with a dynamic viscometer (Haake, Polylab System/52p Rheomex). Thermogravimetric analysis (TGA, Du Pont Instruments, 951 Termogr. Analyze, Wilmington, Germany) of the polymeric foam was performed in air at a heating rate of 10 °C/min in order to establish an appropriate heat treatment. X-ray diffraction (XRD, Diffrac 500, Siemens AG, Mannheim, Germany) with CuKα radiation was used to identify the phases of LZSA glass ceramic foam. The compression strength of a set of 6 samples with nominal dimensions of 20 x 12 x 12 mm was determined at room temperature using a universal testing device (Instron, Model 4202, Instron Corp., Canton, MA, USA). The speed of the crosshead was set constant to 1 mm/min. The strut density was determined by He-pycnometer (AccuPyc 1330, Micromeritics, Norcross, GA). The open porosity was derived from the relation between the strut density and the geometrical density.

The cellular morphology of the LZSA glass ceramic foams was examined using an X-ray microtomograph (μ-CT40, Scanco Medical AG, Bassersdorf, Switzerland). A detailed description of foam structure characterization by μ-CT40 was reported by Zeschky et.al. [16].

RESULTS AND DISCUSSION

The TGA analysis of the polyurethane foam revealed two weight loss stages. The first one between 250 and 400 °C, which corresponds to the decomposition of the polymer, resulted in an accentuated weight loss. The second stage between 400 and 600 °C is attributed to the oxidation of carbon derived from the decomposition of the polymers of the foam [15]. The TGA results allows the choice of the optimized thermal cycle for the decomposition of the foams.

An adequate slurry to impregnate polymeric foams must be fluid enough to enter, fill and coat uniformly the foam network and present enough viscosity under static condition to remain in the foam during drying [5,9]. It is suggested that the slurry must exhibit a thixotropic behavior, which is commonly observed in pseudoplastic slurries [17].

Figure 1 show flow curves of the LZSA parent glass ceramic slurries with different concentrations of bentonite in two different solvents, water (Fig. 1a) and isopropanol (Fig. 1b). Thixotropic fluids with relatively high viscosity may lead to an optimized impregnation. In Fig 1a it can be noticed that the only thixotropic aqueous slurry is that with 1.25wt% of bentonite. This slurry, however, is characterized by too low a viscosity for impregnation. Additionally, due to the poor wettability onto the polyurethane foams [18], the relatively low viscosity of the aqueous slurries caused a rapid flow after impregnation and hindered an homogeneous impregnation. The slurries prepared in isopropanol media are characterized by a pseudoplastic behavior (Fig. 1b). The suspensions prepared with 10% and 5% bentonite exhibit an antithixotropic (or rheopexic) behavior while those prepared with 2.5% and 1.25% bentonite show typical thixotropy. The thixotropy degree of increases as the bentonite load decreases. Higher bentonite loads may induce stronger interactions between the particles in the suspensions, which might cause this rheopexic behavior. Although the slurry prepared with 5% bentonite did not show a thixotropic behaviour, it exhibited an adequate viscosity and a pseudoplastic behavior to coat uniformly the strut network of the foam. It was also observed that the suspensions prepared with isopropanol were stable for a longer time without visible particle sedimentation.

For a large variety of ceramic slurries an appropriate model to describe the rheological properties can be chosen. A rheological model with a yield stress (τ_Y) allows the analysis over a wide range of viscoplastic fluids. Thus, LZSA parent glass ceramics slurries with high solid loads can be well described by the Herschel-Bulkley model, Equation 1 [19]:

$$\tau = \tau_Y + \eta\gamma^n \qquad (1)$$

where τ is the shear stress, τ_Y is the Herschel-Bulkley yield stress, η is the consistency index, γ is the velocity gradient and n is the shear thinning constant.

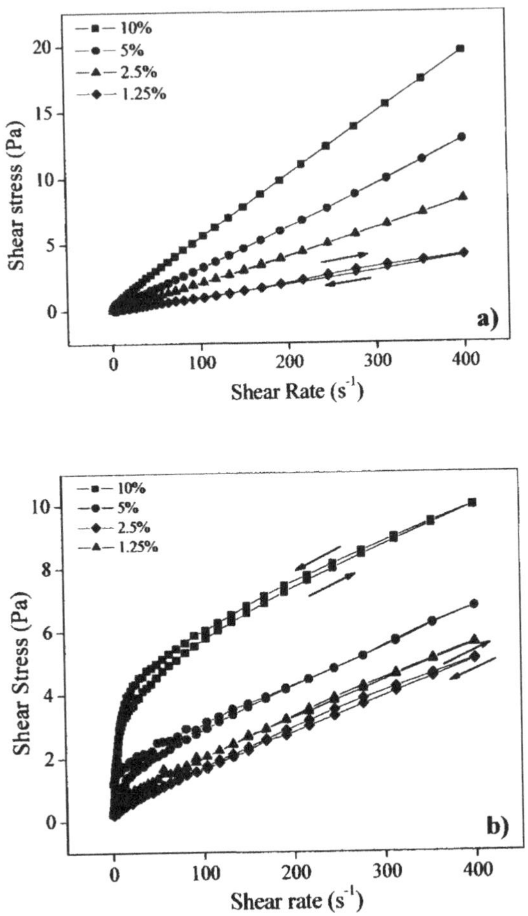

Figure 1: Shear stress *versus* shear rate curves of the LZSA parent glass ceramic slurries, with different concentrations of bentonite, prepared in: a) water; b) isopropanol.

Concentrated suspensions of solid particles in Newtonian liquids often show a yield stress followed by a nearly Newtonian flow [20] as observed in the fitted curves on Fig. 2. It can be seen that the aqueous slurry flows nearly like a Newtonian fluid, even at lower shear rates. The organic slurry, however exhibit an accentuated raise in the shear stress at lower shear rates (< 20 s^{-1}) in the presence of yield stress.

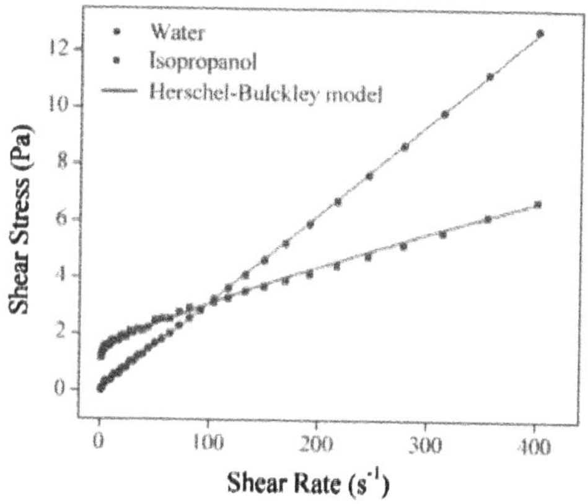

Figure 2: Shear stress versus shear rate plots for the LZSA glass-ceramic slurry in water (•) and isopropanol (•) with 5% bentonite. The data were fitted using the Herschel-Bulckley model (solid curves).

The fitting parameters of the Herschel-Bulckley curves for the LZSA glass ceramic slurries with different amounts of bentonite (binder) in water and isopropanol are shown in Table II.

Table II: Fitting parameters of Herschel-Bulckley (HB) model for the composite LZSA glass ceramic slurries with different amounts of bentonite (binder) in isopropanol and water. The numbers in parenthesis indicates the error of the last digits.

Solvent	wt% bentonite	η (Pa s)	τ_Y (Pa)	n
Water	1.25	0.0078 (5)	0.0174 (8)	1.037 (1)
	2.5	0.0156 (5)	0.093 (8)	1.045 (5)
	5	0.0253 (1)	0.120 (13)	1.036 (6)
	10	0.064 (1)	0.177 (1)	0.951 (4)
Isopropanol	1.25	0.024 (1)	0.235 (8)	0.880 (7)
	2.5	0.0393 (4)	0.403 (23)	0.811 (17)
	5	0.0434 (6)	1.311 (33)	0.805 (23)
	10	0.501 (81)	1.3776 (967)	0.470 (98)

The aqueous slurries exhibit a quasi-Newtonian behavior, characterized by n close to 1. These slurries also exhibit lower viscosities, which is undesirable for the impregnation process. A low viscosity may hinder a uniform coating of the foam, leading to a large number of filled cells after sintering.

For slurries prepared in isopropanol, with 1.25-5 wt% bentonite, n is lower than 1 and nearly constant (less than 10% variation), which indicates shear-thinning with the applied shear stress. A strong decrease in n is observed for the slurry prepared with 10% bentonite, indicating a higher deviation from the Newtonian regime.

Figure 3 shows XRD pattern from the LZSA foam sintered at 900°C for 10 min. The reflections associated with the sintered sample were assigned to the crystalline phases of zirconium silicate (($ZrSiO_4$, JCPDS 6-266), lithium metasilicate (Li_2SiO_3, JCPDS 29-828) and β-spodumene ($LiAlSi_3O_8$, JCPDS 21-503). It can be observed that the major crystalline phases are β-spodumene and zirconium silicate. The β-spodumene phase shows a low coefficient of thermal expansion with a value very close to zero (0.4–2 x 10^{-6}°C^{-1} for β-spodumene based glass-ceramics [15]). Although zirconium silicate shows a higher coefficient of thermal expansion (4.0 x 10^{-6} °C^{-1}), this phase contributes to a high abrasion and chemical resistance of the material. Lithium metasilicate was detected as minor phase.

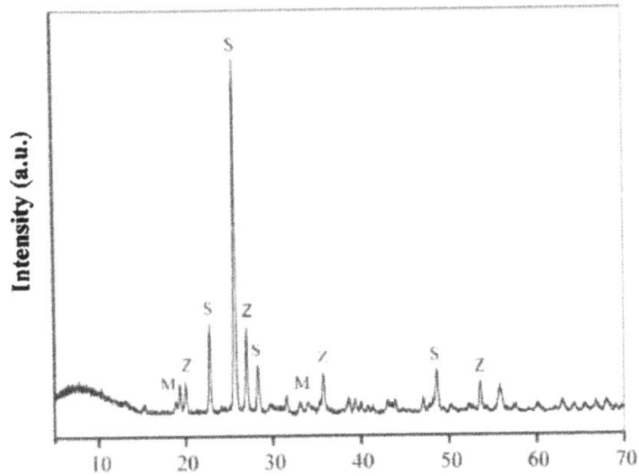

Figure 3: XRD patterns of the LZSA foams sintered at 900°C for 10 min. M = lithium metasilicate; Z = zirconium silicate and S = β-spodumene.

SEM-micrographs of the LZSA foams are shown in Figure 4. It can be observed that the sintered LZSA foams prepared from both aqueous (Fig. 4b) and organic slurries (Figure 4c) maintained the pore structure of the original polyurethane template (Fig. 4a). The pores of the LZSA foams are mostly interconnected and round-shaped. Some closed cells on the surface can also be observed on both samples. Additionally, the foam impregnated with aqueous slurry exhibits a higher inhomogeneity of the pore morphology, characterized by cracks, pore wall defects and cell wall discontinuity (Fig. 4b). Large voids (up to 2 mm) can be seen between the small pores (~0.5 mm) due to this discontinuity forming a mixed pore network These microstructural defects were generated during the impregnation process. As discussed before the poor wettability of the polyurethane foam to water led to a inhomogeneous coating layer, which could be the main reason for these defects.

Therefore, the mechanical strength of the glass-ceramic foams prepared with aqueous slurries will be negatively affected by these defects. On the other hand, the foam impregnated with the organic slurry shows a well defined rounded pore structure with a continuous glass-ceramic strut network (Fig. 4c). A quasi-monomodal pore size distribution was confirmed with mean pore size of 0.2 mm for this foam. Due to the rheology differences of the slurries, the foams impregnated with aqueous slurry shrank approximately 11vol% more in comparison to the foams prepared with isopropanol.

Table III summarizes the physical and microstructural properties of the LZSA foams, derived from He-pycnometry and μ-CT analysis. It is observed that the pycnometric density is slightly lower for the foam prepared with aqueous slurry, indicating the presence of closed porosity.

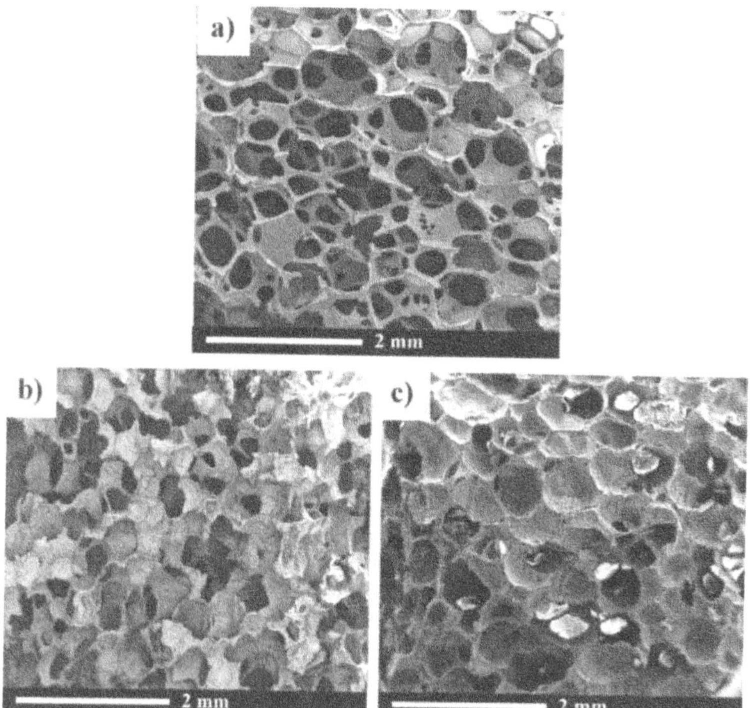

Figure 4: SEM micrographs of the foams. a) PU foam; b) LZSA glass-ceramic foam prepared with aqueous slurry; c) LZSA glass-ceramic foam prepared with organic slurry.

The foams prepared with aqueous slurry exhibit larger cell sizes with a slightly increased number of lager pores. The shrinkage of the foam impregnated with aqueous slurry can be noticeable, based on the struts approximately 30% thinner than the struts from the foam impregnated with organic slurry. The struts of the foams prepared in isopropanol are thicker.

Table III: Physical and morphological properties of the LZSA foams.

	ρ_{rel} (g/cm^3)	ρ_{strut} (g/cm^3)	Porosity (%)	MCS (mm)	MST (mm)
LZSA-water	0.26 ± 0.02	2.67 ± 0.01	90 ± 1	0.30 ± 0.02	0.14 ± 0.02
LZSA-isoprop.	0.42 ± 0.02	2.65 ± 0.01	84 ± 1	0.22 ± 0.02	0.20 ± 0.02

ρ_{rel}=relative density, ρ_{strut}=strut density, MCS=Mean Cell Size, MST=Mean Strut Thickness

The μ-CT results are in good agreement with the SEM observations, which certifies this method to obtain morphological parameters that may be used to control the manufacturing process and allow the microstructure design of the final glass-ceramic foams.

The mean compressive strength was 1.06 ± 0.01 MPa for the LZSA foam prepared with isopropanol and 0.49 ± 0.01 MPa [15] for the LZSA foam prepared using water as slurry solvent. The struts of the foam prepared with organic slurry are thicker, leading to a higher mechanical strength. The decrease in the strength of the foam prepared with aqueous slurry is attributed to the increase in its total porosity, reflecting the inhomogeneity of the particle coating during impregnation, as expected. A more detailed mechanical analysis will be evaluated in the forthcoming work.

CONCLUSIONS

LZSA foams with interconnected pores and high porosity (90-95%) were obtained by the polymeric foam method. The glass-ceramic foams consist of β-spodumene, zircon and lithium metasilicate as the main crystalline phases. The rheological evaluation revealed a quasi-Newtonian behavior of the slurries prepared with different concentrations of bentonite in water. Slurries prepared in isopropanol showed a shear thinning behavior with yield stress. Besides, the organic slurries were characterized by time dependent behavior closely related to the bentonite concentration. The microstructure of the final foams was directly affected by the solvent used to prepare the slurries. Due to inhomogeneous impregnation, microstructure defects and an increase in the porosity were observed in the foams prepared with aqueous slurry, which caused a decreased mechanical strength under compressive load.

ACKNOWLEDGEMENTS

The authors are thankful to Prof. Peter Greil for many helpful discussions. The authors also thank CAPES/Brazil, CNPq/Brazil and to the Government of Bavaria/Germany for funding this work under Grant No. 007/02.

REFERENCES

[1]P. Colombo, and E. Bernardo, "Macro-and Micro-Cellular Porous Ceramics from Preceramic Polymers," *Composites Sci. Tech.*, **63**, 2353–59 (2003).

[2]P. Ciambelli, V. Palma, P. Russo, and S. Vaccaro, "Performances of a Catalytic Foam Trap for Soot Batement," *Catalysis Today*, **75**, 471–78 (2002).

[3]M.D.M. Innocentini, A.R.F. Pardo, B.A. Menegazzo, L.R.M. Bittencourt, R.P. Rettore, and V.C. Pandolfelli, "Permeability of High-alumina Refractory Castables Based on Various

Hydraulic Binders. *J. Am. Ceram. Soc.*, **85**, 1517–21 (2002).

[4]P. Sepulveda, "Gelcasting Foams for Porous Ceramics", *Am. Ceram. Soc. Bull.*, **76**, 61-65 (1997).

[5]L. Montanaro, Y. Jorand, G. Fantozzi, A. Negro, "Ceramic Foams by Powder Processing", *J. Eur. Ceram. Soc.*, **18**, 1339-50 (1998).

[6] R. Brezny, and D.J Green., "Mechanical Behaviour of Celular Solids". In *Materials Science and Technology*, **11**, Structure and Properties of Ceramics, ed. R.W. Cahn, P Haasen and E. J. Kramer. VCH, Germany, 467-516 (1992).

[7]L.J. Gibson and M.F. Ashby, "Cellular Solids: Structure and Properties", Pergamon Press, New York, 1988.

[8]K. Schwartzwalder, A.V. Somers, "Method of making porous ceramics articles", *US Pat.* No. 3 090 094, May 21, 1963.

[9]J. Saggio-Woyansky, C.E. Scottetal, "Processing of Porous Ceramics", *Amer. Ceram. Soc. Bull.*, **71**, 1674-82 (1992).

[10]P. Sepulveda, J.G.P. Binner, "Processing of Cellular Ceramics by Foaming and in situ Polymerisation of Organic Monomers", *J. Eur. Ceram. Soc.*, **19**, 2059-66 (1999).

[11]H.X. Peng, Z. Fan, J.R.G. Evans, J.J.C. Busfield, "Microstructure of Ceramic Foams", *J. Eur. Ceram. Soc.*, **20**, 807-13 (2000).

[12]T.J. Fiztgerald, A. Mortensen, V.J. Michaud, "Processing of Micro-Cellular SiC Foams. Part II: Ceramic Foam Production", *Journal of Materials Science*, **30**, 1037-45 (1995).

[13]N.M. Bobkova, N.I. Zayatz, T.V. Kolontaeva, G.N. Punko, G.B. Zakharevich, "Porous Glass Ceramic Bioimplants", *Glass and ceramics*, **57**, 412-14 (2000).

[14]O. R. K. Montedo, A. P. Novaes de Oliveira, A. N. Klein, PTECH 2001 – *3rd Latin-American Conf. on Powder Technology*, Florianópolis, SC, Brazil, November 2001.

[15]E. Sousa, C.B. Silveira, T. Fey, P. Greil, D. Hotza, A.P.N. Oliveira, "LZSA Glass Ceramic Foams Prepared by Replication Process", *Advances In Applied Ceramics*, **104**, 22-9 (2005).

[16]J. Zeschky, T. Höfner, C. Arnold, R. Weißmann, D. Bahloul-Hourlier, M. Scheffler, P. Greil, "Polysilsesquioxane Derived Ceramic Foams with Gradient Porosity", *Acta Materialia*, **53**, 927-37 (2005).

[17]X. Zhu, D. Jiang, S. Tan "The Control of Slurry Rheological in the Processing of Reticulated Porous Ceramics", *Materials Research Bulletin*, **37**, 541-53 (2002).

[18]S.K. Koh, J.S. Cho, K.H. Kim, S. Han, Y.W. Beag, "Altering a Polymer Surface Chemical Structure by an Ion-assisted Reaction", *Journal of Adhesion Science and Technology*, **16**, 129-42 (2002).

[19]J.S. Reed, "Principles of Ceramic Processing", Wiley-Interscience, New York, 1995, p. 277-309

[20]C.W. Macosko, "Rheology Principles, Measurements, and Applications", VHC Publisher, Inc. 1994.

Eco-Materials Processing and Design
Edited by Koji Watari and Soo-Wohn Lee

LOW-ENERGY SYNTHESIS OF SOLID OXIDE FUEL CELL BY MICRO-WAVE SINTERING

S.Takahashi, S.Suda , K.Jono, K.Kawahara
Japan Fine Ceramics Center
2-4-1 Mutsuno, Atsuta-ku
Nagoya, 456-8587 Japan

A.Doi
DAIDO Institute of Technology
10-3 Takiharu-cho, Minami-ku
Nagoya, 457-8530 Japan

ABSTRACT

The densification behavior of 8mol%Y_2O_3 doped ZrO_2 (8YSZ) by microwave sintering was studied. 8YSZ dense electrolytes (relative density > 98%) were obtained by microwave dielectric sintering at 1150°C and higher for 10 min in air, and the sintering led to lower sintering temperature by 250°C compared with conventional method. The dense 8YSZ ceramics composed with small and controlled size grains were prepared by the microwave sintering.

INTRODUCTION

Solid oxide fuel cell (SOFC) is one of the hopeful systems, because it can efficiently convert chemical energy to electricity and can use waste heat and exhaust no CO_2 and work with low noise. SOFC is composed of electrolyte, anode and cathode. In cathode, oxygen molecules are ionized to be oxide ions. The oxide ions pass thorough the electrolyte and reach to the anode. In anode, the oxide ions react with hydrogen molecules to H_2O. In order to improve the performance of the cells, works about the anode-supported cells has been carried out. The components were required to sinter at near temperatures in order to fabricate devices with such a complex layered structure. The sintering temperatures of the usual electrolytes were higher of 200 to 300°C than those of the anode or cathode. Therefore, the development of the technique to lower the sintering temperature was desired.

Microwave sintering was studied as one of the method to solve the problem. Microwave energy was directly absorbed by materials with a higher dielectric loss. Therefore the microwave was considered to heat more effectively the materials than conventional method. The microwave energy P absorbed into a dielectric material with volume V_s is estimated to be

$$P = 1/2\varepsilon_0\varepsilon''\omega E^2 V_s$$

where ε_0 is the dielectric constant in a vacuum, ε" is the dielectric loss, ω is the angular frequency and E is the electrical field strength. The ε"s of the electrolytes which are typically yttrium stabilized zirconia, (La,Sr)(Ga,Mn)O_3, etc. are ordinarily higher than those of electrodes. This means the electrolyte can be sintered at lower temperature by microwave sintering than by using normal electric furnace and the difference of the sintering temperatures between the electrolyte and the electrodes decreases.

There are several studies about microwave sintering of zirconia based ceramics [1-6]. These reported this method was effective to decrease the sintering temperature. It was worthy of note that M.A.Janney et al.[6] succeeded to sinter the zirconia at about 1200°C whereas it can be sintered at 1350°C using a electric furnace.

Another advantage of the microwave is to be possible to sinter rapidly because it can directly heat the materials. At first, the insulating material of the furnace was heated and then the samples were heated by heat transfer in the conventional sintering. The total energy was expected to be decreased by microwave sintering because it was the product of the sintering time and the electric power which was guessed to depend on the sintering temperature.

In this report, we discussed about the densification behavior of 8mol%Y_2O_3 doped ZrO_2 by microwave sintering and tried to fabricate a single cell of the SOFC by the method preliminarily.

EXPERIMENTAL PROCEDURE

All samples were prepared from 8mol% yttria – zirconia powder (Tosoh). The powder was poured into a 18mm diameter cylindrical isopress mold. The rods were pressed and discs were taken out and CIPed. The zirconia disks set in a furnace center. The disk was surrounded by SiC bricks which were good absorbers of microwave and assistants for heating. The sample temperature was measured with radiation thermometers through an upper hole of the sample room. Power was supplied by 2kW, continuously variable 2.45 GHz microwave generator. The microwave power was controlled through feedback control from a proportional–integral-derivative controller. Heating ratio was 40°C /min, heating time was 10 min. and sintering temperatures were 1100-1400°C and atmosphere was air. The samples were sintered in a conventional electric furnace at the same conditions for comparing with samples sintered by microwave sintering. Relative density of the obtained samples was measured and the microstructure was observed by SEM.

RESULTS AND DISSCUSSION

Densification of 8YSZ disks

The relative density and grain size of the samples sintered by microwave and conventional method were shown in Figure1 and 2, respectively. The SEM photographs of the surface of the samples were shown in Figure 3. The relative density of the sample sintered by microwave at 1100°C was about 90% and at higher than 1150°C was more than 98%. The density by conventional method at 1300°C was about 83% and at 1400°C and higher was more than 98%. From the results, it was found that the microwave sintering decreased the sintering temperature by about 250°C compared with conventional method. The result was better than those in reference 6 from the viewpoint of low energy sintering. This was probably caused by a good matching between microwave power and the arrangement of the sample and assistants.

The grain size of the samples sintered by electric furnace quickly increased at 1400°C and higher. On the other hand, the grain size of the samples by microwave slowly increased from 0.3 to 1.6μm as the sintering temperature increased. The result seems to be useful for preparing the dense zirconia ceramics composed with small size grains.

When the used sintering energy was supposed to depend on a product of the sintering temperature and time, the energy by microwave sintering was 78% of that by electric furnace. (When the heating ratio was 10°C /min in conventional method , the energy by microwave sintering decreased to 27% of that by electric furnace.)

SOFC Preparation

SOFC samples was prepared by microwave sintering using NiO-8YSZ as anode and $La_{0.8}Sr_{0.2}MnO_3$(LSM□as cathode. The SEM photographs of broken surface of the samples were shown in Figure 4. There was many porosities in the electrolyte of the samples sintered at 1100°C by microwave, whereas the electrolyte parts sintered at 1150 and 1200°C were dense. This result agreed with that of the disk ceramics. The interface between the electrolyte and anode of the sample sintered at 1150°C was clear and smooth, whereas that of the sample sintered at 1400°C by electric furnace was rough. This was probably caused by the grain growth was depress by microwave sintering.

SUMMARY

Microwave sintering was tried for densification of electrolytes for SOFC. The results were as follows.

1) The 8YSZ ceramics with relative density of 98% and more was prepared at 1150°C for 10 min. by microwave sintering.
2) The microwave sintering decreased the sintering temperature by about 250°C compared with conventional method.
3) The dense zirconia ceramics composed with small and controlled size grains were prepared by microwave sintering.

REFERENCES

[1]S. A. Nightingale, H. K. Worner, and D. P. Dunne, "Microstructural Development during the Microwave Sintering of Yttria-Zirconia Ceramics," *J. Am. Ceram. Soc.*, **80**, 394-400 (1997).

[2]Y. Fang, J. Cheng, R. Roy, D. M. Roy, D. K. Agrawal, "Enhancing densification of zirconia-containing ceramic-matrix composites by microwave processing," *J. Mater. Sci.*, **32**, 4925-4930 (1997).

[3]S. Fujitsu, M. Ikegami and T. Hayashi, "Sintering of Partially Stabilized Zirconia by Microwave Heating Using $ZnO-MnO_2-Al_2O_3$ Plates in a Domestic Microwave Oven," *J. Am. Ceram. Soc.*, **83**, 2085-87 (2000).

[4]A. Kishimoto, M. Ito, S. Fujitsu, "Microwave sintering of ion conductive zirconia based composite dispersed with alumina," *J. Mater. Sci. Lett.*, **20**, 943-945 (2001).

[5]D. D. Upadhyaya, A. Ghosh, G. K. Dey, R. Prasad, A. K. Suri, "Microwave sintering of zirconia ceramics," *J. Mater. Sci.*, **36**, 4707-4710 (2001).

[6]M. A. Janney, C. L. Calhoun, H. D. Kimrey, "Microwave sintering of Solid Oxide Fuel Cell Materials: 1, Zirconia-8mol% Yttria," *J. Am. Ceram. Soc.*, **75**, 341-46 (1992).

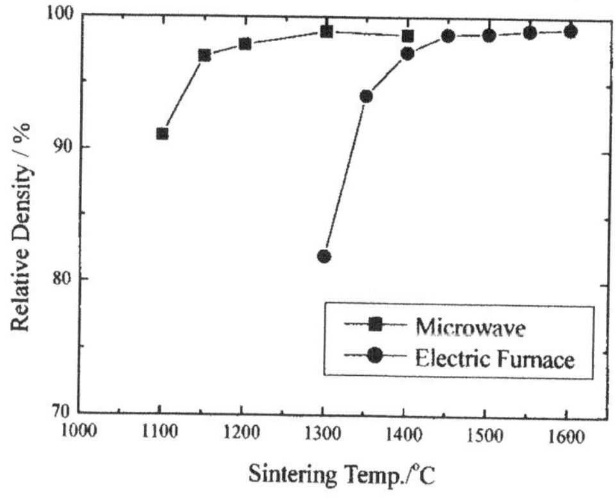

Figure 1 Relationship between relative density of 8YSZ ceramic and sintering temperatures.

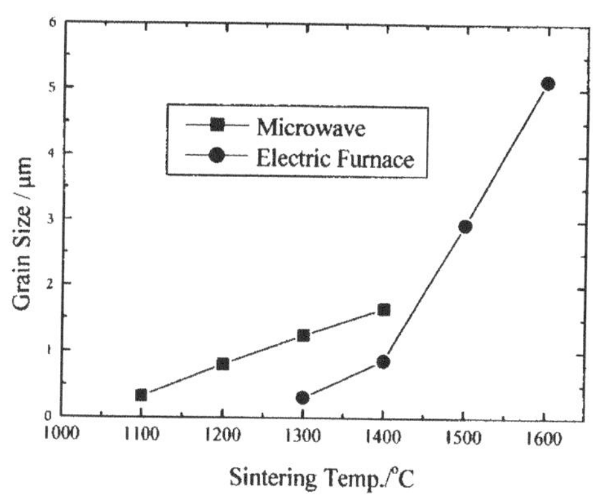

Figure 2 Relationship between grain size of 8YSZ ceramic and sintering temperatures.

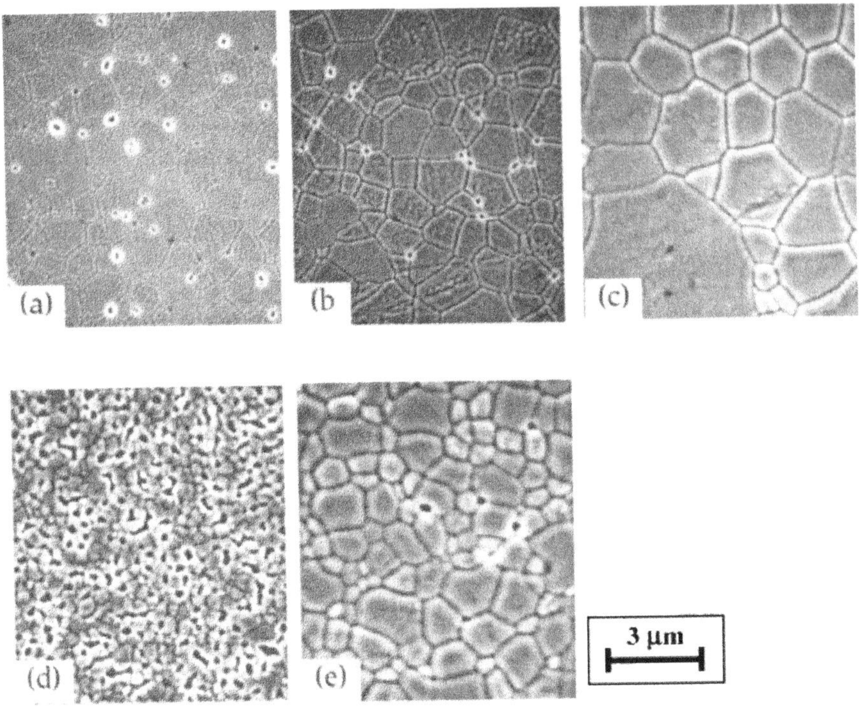

Figure 3 SEM photographs of the surface of the samples sintered at (a)1200 (b)1300 (c)1400°C by microwave sintering, at (d) 1300 (e)1400°C using conventional electric furnace.

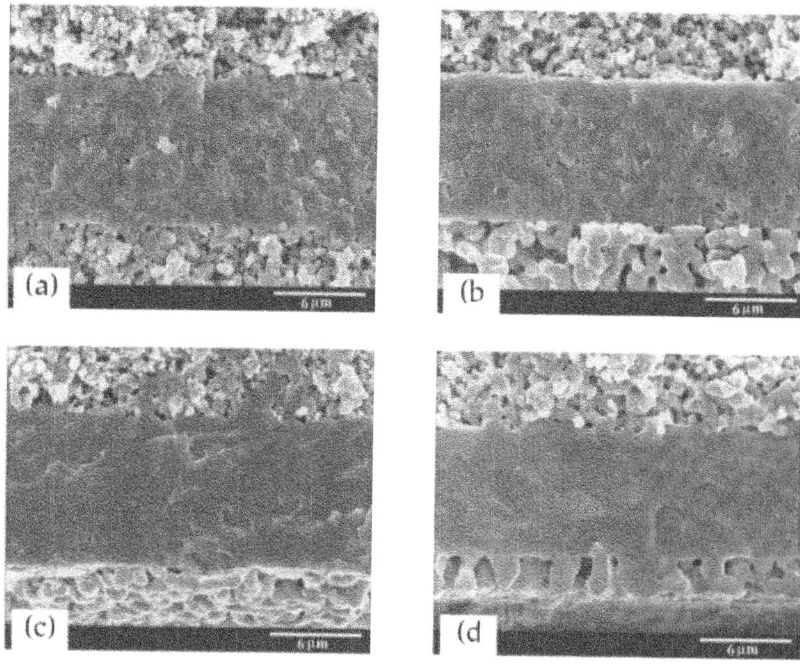

Figure 4 SEM photographs of the broken surface of the cells sintered at (a)1100 (b)1150 (c)1200°C by microwave sintering, at (d) 1400°C using conventional electric furnace. Upper side is NiO-8YSZ, middle part is 8YSZ and lower side is $La_{0.8}Sr_{0.2}MnO_3$.

Eco-Materials Processing and Design
Edited by Koji Watari and Soo-Wohn Lee

REDUCTION OF CARBON DIOXIDES EMISSION IN THE FERROCHROME MANUFACTURING WITH ALUMINUM

Hideya Matsumura
Major of Material Systems Engineering and Life Science, Graduate School of Science and Engineering, Toyama University, 3190 Gofuku, Toyama, 930-8555, Japan

Takashi Hashizume, Atsushi Saiki and Kiyoshi Terayama
Department of Material Systems Engineering and Life Science, Faculty of Engineering, Toyama University, 3190 Gofuku, Toyama, 930-8555, Japan

ABSTRACT

The present study concerns to the thermoanalytical investigation on the kinetics of the reduction process of $FeCr_2O_4$ with alumimum powder under non-isothermal conditions. In the present investigation, simultaneous thermogravimetry (TG) - differencial thermal analysis(DTA) technique was used. SEM and XRD were also used in order to identify the phases in the samples quenched at several chosen conditions.

On the DTA curve, three peaks were observed corresponding to the endothermic peak for melting of aluminum, the exothermic heat change for oxidation of excess aluminum, and heat change for reoxidation of ferrochrome, respectively.

The analysis of the product indicated the presence of Al_2O_3, Cr_2O_3, Fe-Cr and unreacted Al. After that, Fe-Cr and unreacted Al were oxidized gradually.

INTRODUCTION

Although carbothermic process is the most economical and widely used refining process for ferroalloys, several carbides are formed evolving carbon dioxide. In our laboratory, the carbothermic reductions of Cr_2O_3 and $FeCr_2O_4$ were investigated with thermally evolved gas analysis method (EGA). Carbides of Cr_3C_2 and $(Fe, Cr)_7C_3$ were formed as a reduction products as well as the case of oxide [1]. It became clear that the reduction was much influenced by the carbide formation. This problems arising out of using carbon as a reductant can be overcome by using aluminum as a reductant. On the other hand, aluminum being a costly reducer, it is very much important to know the optimum aluminum requirement for maximum recovery of alloys [2, 3]. The kinetics of the carbothermic reduction mechanism of Cr_2O_3 and $FeCr_2O_4$ has been investigated and the reaction models were successfully proposed.

As ferrochrome of high purity can be obtained with aluminum, the reduction mechanism of $FeCr_2O_4$ with aluminum is not clear. The present work aims to clear the aluminothermic reduction process of $FeCr_2O_4$, as well as to reduce the emission of carbon dioxide and energy consumption through the processing. Simultaneous TG-DTA technique was used. Samples quenched at several chosen conditions were used for identifying the phases by means of SEM and XRD.

EXPERIMENTAL

The TG-DTA experiments were carried out in TG-DTA 2000S model of Bruker AXS. The reactant mixtures were composed of powder of $FeCr_2O_4$ and powder of Al (99.5wt% purity) with several molar ratio.

In each run a sample weighing 10mg was put into alumina cylindrical pan and heated up to 1473K in a dynamic atmosphere of air of 100ml/min. Alumina powder was used as reference material for DTA.

The reaction products have been detected by X-ray diffraction technique.

RESULTS AND DISCUSSION

Kinetic study of reduction of $FeCr_2O_4$ with Al has been carried out with molar ratio $FeCr_2O_4$: Al=1.5:10.

The typical reduction process of $FeCr_2O_4$ under non-isothermal condition was shown in Fig.1. Three peaks were observed on the DTA curve corresponding to, a) the endothermic peak for melting of aluminum, b) the exothermic heat change for oxidation of excess aluminum, and c) heat change for reoxidation of ferrochrome, respectively.

On the other hand, the mass gain occured gradually at about 700 K, and the rate of mass gain increased as the temperature increased from 1100 K, which correspond to those of oxidation of the part of the excess aluminum in the mixture.

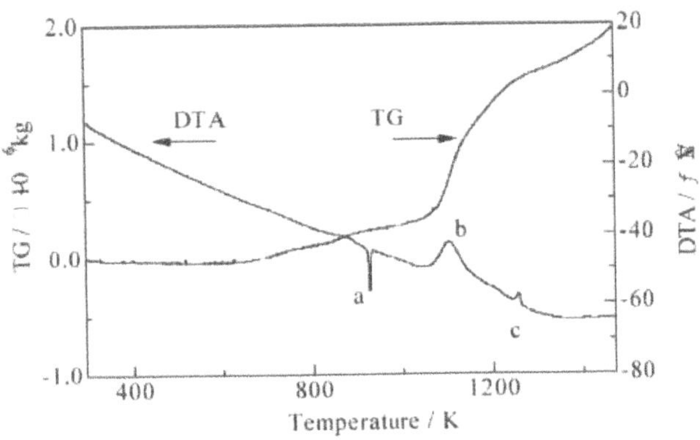

Fig. 1. TG and DTA curves for the mixture with molar ratio $FeCr_2O_4$:Al=1.5:10

CONCLUSION

The following conclusions have been obtained from the present study:

(1) The reduction seemed not to proceed under $FeCr_2O_4$: Al=1.5:10 molar ratio.

(2) The analysis of the product indicated the presence of Al_2O_3, Cr_2O_3, Fe-Cr and unreacted Al. After that, Fe-Cr and unreacted Al were oxidized gradually to Fe_2O_3, Cr_2O_3 and Al_2O_3.

(3) It is possible to obtain high purity Fe-Cr by aluminothermic reduction.

The application of aluminothermic reduction technique to the ferroalloy making process as well as the ceramic processing, will contribute toward reducing the emission of carbon dioxide and the other poisonous gases, becoming a world-wide problem.

REFERENCES

[1] H.Hirazawa, T.Hashizume, A.Saiki and K.Terayama, "REDUCTION ANALYSIS BETWEEN CHROMIUM CARBIDE AND CHROMIUM OXIDE BY THERMALLY EVOLVED GAS ANALYSIS, " ICCT 2004, Beijing, China,pp.122.

[2] B.SARANGI, A.SARANGI and H.S.RAY, "Kinetics of Aluminothermic Reduction of MnO_2 and Fe_2O_3 : Al Thermoanalytical Investigation," ISIJ International, Vol.36(1996),No.9,pp.1135-1141

[3] Osvaldo Mitsuyuki CINTHO, Claudio Parra De LAZZARI and Jose Deodoro Trani CAPOCCHI, "Kinetics of the Non-isothermal Reduction of Cr_2O_3 with Aluminium," ISIJ International, Vol. 44(2004), No.5, pp.781-784

Eco-Materials Processing and Design
Edited by Koji Watari and Soo-Wohn Lee

THE EFFECT OF PRETREATMENT ON HYDROTHERMAL PROCESSING FOR RECYCLING OF ZIRCONIA PRODUCT

Tomohiko OGATA[1], Takuya MASTUBARA[1], Hiromi NAKANO[2], Kazuyori URABE[2], Toshihiko NISHIDA[3]

1) Ceramics Division, Toray Industries Inc., 3-2-2 Sonoyama, Otsu-shi, Japan, 520-0842

2) Electron Microscope Lab., Faculty of Engineering, Ryukoku University, 1-5, Yokoya, Seta-Ohe-cho, Otsu-shi, Japan, 520-2194

3) Department of Chemistry and Materials Technology, Kyoto Institute of Technology, Matsugasaki, Sakyo-ku, Kyoto-shi, Japan, 606-8585

ABSTRACT

We studied the practical application of re-using sintered yttria-stabilized tetragonal zirconia polycrystal (Y-TZP) by hydrothermal processing. The Y-TZP was first cooled rapidly from 1000°C to introduce microscopic cracks into the sintered material, and was then pulverized by hydrothermal processing for 66 hours and 162 hours under hydrothermal conditions using a autoclave at 200°C. The obtained powder was ground down, shaped and sintered, and with each powder it was possible to regenerate a dense-textured Y-TZP sintered material. This preprocessing allows the hydrothermal processing time to be reduced, and should lead to a reduction in processing costs.

INTRODUCTION

Partially stabilized zirconia ceramics (PSZ), which are both strong and tough, have been expected to expand the market as mechanical element materials and in structural applications as a substitute for metallic materials that are impossible to use under excessively harsh conditions. However, PSZ have a fatal defect, degradation on the strength in hydrothermal condition.[1)]

The authors have already reported on the results of their investigation into the possibility of recycling by focusing on the hydrothermal degradation of zirconia ceramics.[2)] Specifically, when partially stabilized zirconia was subjected to hydrothermal processing and pulverized by promoting the degradation of the sintered material entirely, it was possible to use this powder as a raw material for the regeneration of sintered materials again by shaping and sintering it.

However, one of the practical difficulties involved in recycling zirconia sintered materials is that of converting zirconia sintered materials into fine powders quickly and inexpensively. The authors have noticed that to break down the sintered material quickly it is necessary to obtain a powder with as small a primary particle size as possible by subjecting it to treatments such as introducing cracks and making the crystal phases non-uniform to reduce its resistance to hydrothermal corrosion. Specifically, hydrothermal corrosion is thought to occur when OH bonds in water break the bonds between Zr atoms at the surface of the sintered material.[3)] To increase the rate of corrosion, it is therefore necessary to create a large surface area by introducing as many pre-cracks as possible in the sintered material. In this study, we subjected zirconia sintered materials to thermal shock to introduce defects before the hydrothermal processing, and as shown in this report we were able to substantially reduce the hydrothermal processing time compared with the case where no pretreatment was performed, resulting in the discovery of a practical regeneration technique.

EXPERIMENTAL PROCEDURE

PRETREATMENT

The Y-TZP sintered material used in these experiments consisted of unused grinding beads with a diameter of 1.5 mm made by rolling pelletization of a powder produced by the coprecipitation method, followed by sintering for 2 hours at 1400°C. Table 1 lists the additives used in the Y-TZP sintered material, and the levels of impurities within it.

Table 1. Additives and impurities of starting Y-TZP (Toreceram beads 1.5 mmØ).

	Y_2O_3	Al_2O_3	SiO_2	CaO	MgO	Fe_2O_3
mass %	4.96	0.389	0.01	0.0021	0.0002	0.0003

These 1.5 mm diameter beads were placed in a Kanthal Super electric furnace (LHT16/R17 made by Chugai Engineering Co., Ltd.), held at the prescribed temperature for 1 hour, then removed and plunged directly into iced water kept at a temperature of 0°C. The electric furnace was set to three different temperatures: 600°C, 800°C and 1000°C.

The beads that had been heat-treated at each temperature were dipped in a dye penetrant liquid (FAW-3, made by Taseto Co., Ltd.), and were then ground down to the middle to allow the state of pre-cracking to be observed by SEM. Beads that had been processed in the above manner were also subjected to compressive failure testing at a crosshead speed of 0.5 mm/minute in a universal testing machine (CATY2000, made by Yonekura Mfg. Co., Ltd.) to determine the compressive failure load.

HYDROTHERMAL TREATMENT

Beads that had been processed in the above way were weighed out in 500 g batches to which 200 g of distilled ion-exchange water was added, and were then subjected to hydrothermal processing for 66 hours and 162 hours in an autoclave (NU-4, made by Nitto Kouatsu Co., Ltd.) at 200°C. The rotation speed of the autoclave stirring paddle was fixed at 200 rpm.

The hydrothermal processing yielded a cloudy corroded product which was separated into precipitate, suspended particles and supernatant solution as described below.

First, the cloudy liquid produced after processing was stirred, allowed to settle for 30 minutes, and separated into a precipitate and a suspension. The suspension was then recovered by separating it into suspended particles and supernatant liquid with a centrifuge (HIMAC Centrifuge CT5DL, made by Hitachi) running at 1500 rpm for 20 minutes.

POWDER PREPARATION, SHAPING AND SINTERING

The sedimentated powders which were obtained by above operations were pulverized and dispersed by attrition mill (using 3mmØ and 5mmØzirconia beads blended). These powders were pulverized for 2h, added 1.2l. of ion exchange water to 500g of powder materials, and were collected through 32 mesh of stainless sieve.

When these collected powders were dried, over 90% of the starting beads were recoverable as recycled powders.

These recycled powders were shaped by die pressing at 29MPa without binders, and were compacted by CIP (made by Mitsubishi heavy industries Co.) at 98MPa. The final densities of them were about 2.95 g/cm3.

These green blocks were sintered in the electric furnace (LHT16/R17, made by Chugai Eng. Co.) at 1400, 1450 and 1500°C for 2h.

CHARACTERIZATION

The characterization items of powder mentioned above were carried out as follows.

Specific surface area measurement by BET machine (made by Quanta Chrome, MS-17), particle size distribution by Laser scattering method (made by Horiba manufacturing Co., LA-920) and identification of crystalline structure by XRD (Rigaku Co., Gygerflex CN4037A1, 40kV, 20mA) were performed.

The content ratio of monoclinic phase in Y-TZP (V_m) was calculated by the following formula (1).

$$V_m(vol\%)=(I_{M(111)}+I_{M(11\text{-}1)})/(I_{M(111)}+I_{M(11\text{-}1)}+I_{T(111)})X100 \qquad (1)$$

Here, $I_{M(111)}$: integral intensity of 111 deflection of monoclinic zirconia

- $I_{M(11\text{-}1)}$: integral intensity of 11-1 deflection of monoclinic zirconia

  $I_{T(111)}$: integral intensity of 111 deflection of tetragonal zirconia

Yttria content in powders was measured by ICP emission spectroscopy after dissolved in sulfuric acid. And the morphology of powder and agglomerate conditions were observed by SEM (JEM-5410, made by JEOL).

As characterization items of sintered Y-TZP, we carried out density measurement by Archimedes method, three pointed bending test based on JIS R 1601, Vickers hardness measurement based on JIS R 1610 and fracture toughness measurement by MI method. Furthermore, the surfaces of Y-TZP were polished, thermal etched (1350°C, 2 h) and observed by SEM (ESA-2000SYSTEM II, made by Elionix Co.) to investigate microstructure of it.

RESULTS AND DISCUSSIONS

RESULTS OF PRETREATMENT

Figure 1 shows the results of performing preprocessing to introduce pre-cracks into the zirconia sintered material. Although mechanical strength of TZP is very high, a heat conductivity coefficient of it is low, and TZP is comparatively weak to a thermal shock.

Heating temperature (to quench)			
blank	600°C	800°C	1000°C
Precracked conditions by dye penetrant liquid			

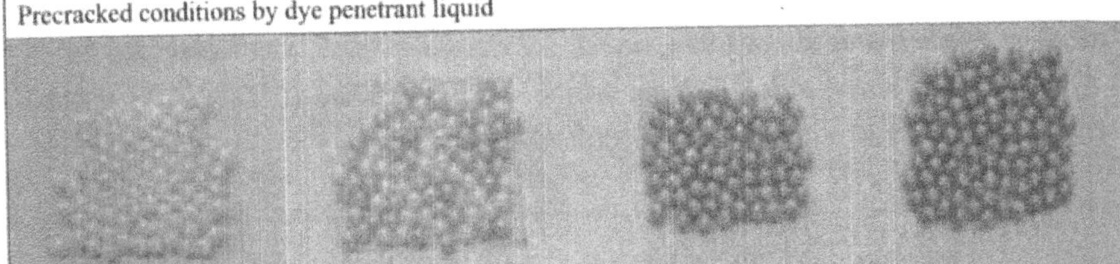

Fracture load by compressive failure testing (N)			
1370 ± 252 (Min1098 ~ Max1784)	754 ± 668 (Min39 ~ Max1784)	122 ± 40 (Min59 ~ Max196)	110 ± 28 (Min39 ~ Max137)

Fig. 1 Precracked conditions after thermal shock pretreatment.

As the results clearly show, the beads subjected to a thermal shock from 600°C remained partially free of cracks, while subjecting the beads to a thermal shock from

800°C or above made it possible to introduce cracks over almost the entire bead surface. This finding is also clearly supported by the results of the compressive failure loading tests. These results demonstrate that a 600°C thermal shock leaves almost all of the beads free from substantial damage, while a thermal shock of 800°C or more causes the beads to be damaged uniformly.

These beads were set in resin and ground down to the middle to allow the state of crack propagation to be observed. Figure 2 shows how the cracks progress in a circumferential direction from the surface to the center. The generating state of cracks were so intense that the temperature was high. From the above result, we chose 1000°C as a prescribed temperature for thermal shock pretreatment. The reason is that the average intensity of 1000°C sample is lower than that of the sample at 800°C, and more internal cracks of the sample at 1000°C are observed by microscope.

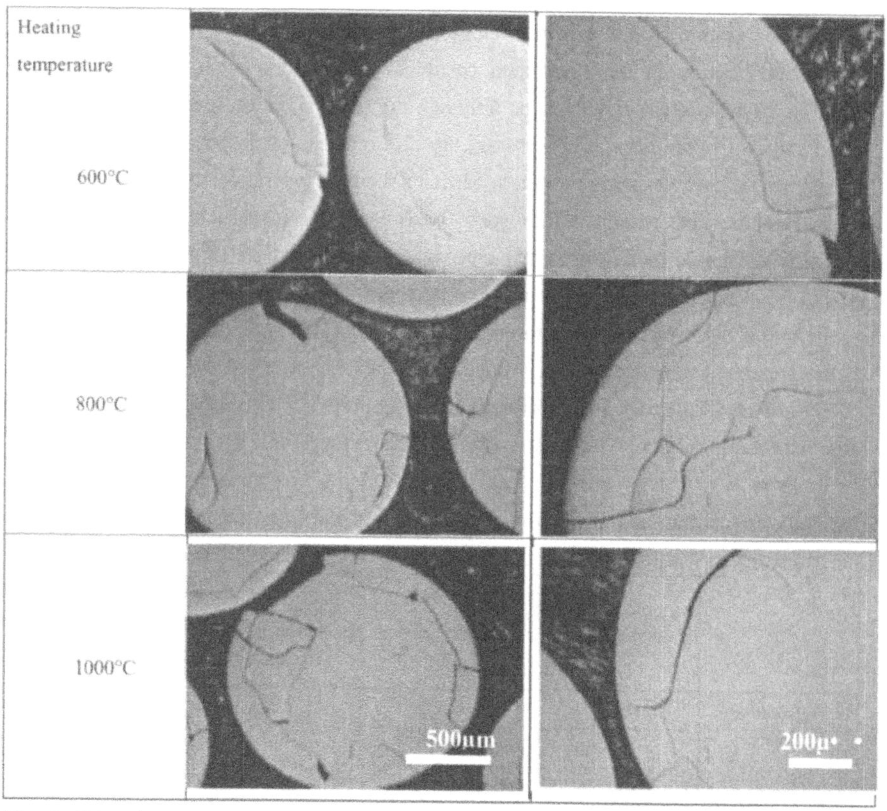

Fig. 2 SEM observation of precracked conditions after thermal shock pretreatment.

The beads used in this study had a diameter of 1.5 mm; for smaller beads, considering the propagation of heat from the surface to the center, it is likely that an even larger temperature difference would be needed.[4)]

CHEMICAL COMPOSITIONS OF Y-TZP AFTER HYDROTHERMAL TREATMENT

Hereafter, we gave a heat shock to beads from 1000°C, and performed hydrothermal processing. The mass of the suspended particles obtained by centrifuging the suspension after autoclave processing was no more than 0.5 g after 66 hours or 162 hours processing, which corresponds to just 0.1 mass% or less of the initial mass.

The authors have already shown that in cases where the beads are not subjected to preprocessing, there is a difference in the yttria content of the precipitate powder and suspended particles after hydrothermal processing for 66 hours, and have reported that this difference might disappear after processing for 162 hours.[2)] This is thought to be because at the initial stage the hydrothermal corrosion proceeds from parts where the yttria concentration is low and the tetragonal crystals have low stability, so that the yttria concentration in the suspended particles is lower than that of the precipitate particles. In the present experiments, although the same situation seemed to arise in the initial stage of hydrothermal processing, the difference in yttria concentration between the suspended particles and precipitate particles would have almost entirely disappeared after 66 hours. This reason is thought to be because the introduction of pre-cracking advanced the grade of hydrothermal corrosion for a short time. If it assumes that the grade of corrosion was the same in hydrothermal treatment of 66 hours and 162 hours by thermal shock treatment, it is thought that early surface area increased at least 140% by this treatment since corrosion advances from the surface.

There was also hardly any difference in the quantity of yttrium ions eluted into the supernatant liquid after 66 hours and after 162 hours (see Table 2).

Table 2. Concentration of yttrium in sedimentated particles, suspended particles and supernatant liquid after hydrothermal treatment.

	After treatment for 66 h		After treatment for 162 h	
	suspended particles	sedimentated particlesr	suspended particles	sedimentated particlesr
Y_2O_3 [mass%]	4.8	4.9	4.8	4.3
	supernatant liquid		supernatant liquid	
Y ion [mg/l]	<0.05		<0.05	

POWDER CHARACTERIZATION AFTER HYDROTHERMAL TREATMENT

We measured the x-ray diffraction profiles of the precipitate powders obtained after 66 hours and 162 hours of hydrothermal processing, which showed that these powders consisted of at least 90vol% monoclinic zirconia. The powder yielded by hydrothermal processing was a mixed phases consisting of tetragonal crystals and monoclinic crystals, making up 90.7vol% of the powder when treatment was performed for 66 hours at 200°C, and 91.3vol% when treatment was performed for 162 hours (see Fig.3). The original sintered material contained only a tetragonal phase and no monoclinic phase was detected, so it seems that the hydrothermal processing caused most of the tetragonal phase to transform into a monoclinic phase.

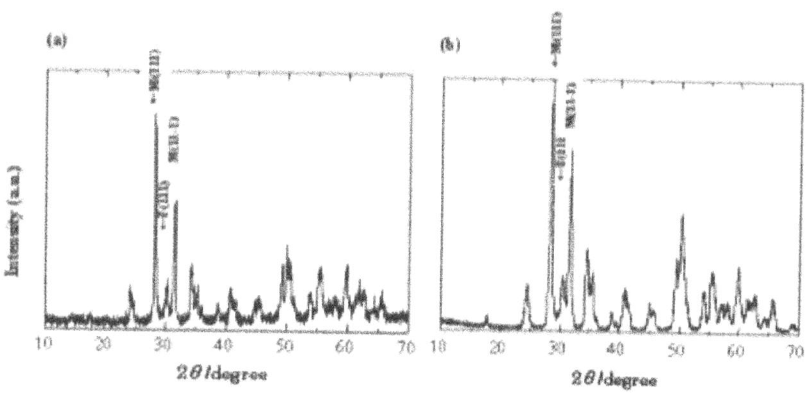

Fig. 3 Powder XRD patterns of sedimentated particles after hydrothermal treatment for (a) 66 h and (b) 162 h.

The BET surface area of the precipitate powders obtained after 66 hours and 162 hours of hydrothermal processing was 3.5 m^2/g in both cases, and their mean aggregate sizes were 0.65 μm and 0.60 μm respectively. The particle size distribution histograms had a single peak (see Fig. 4).

(a)

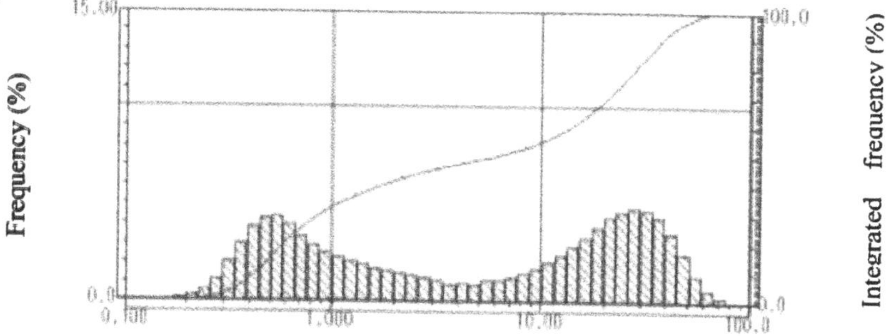

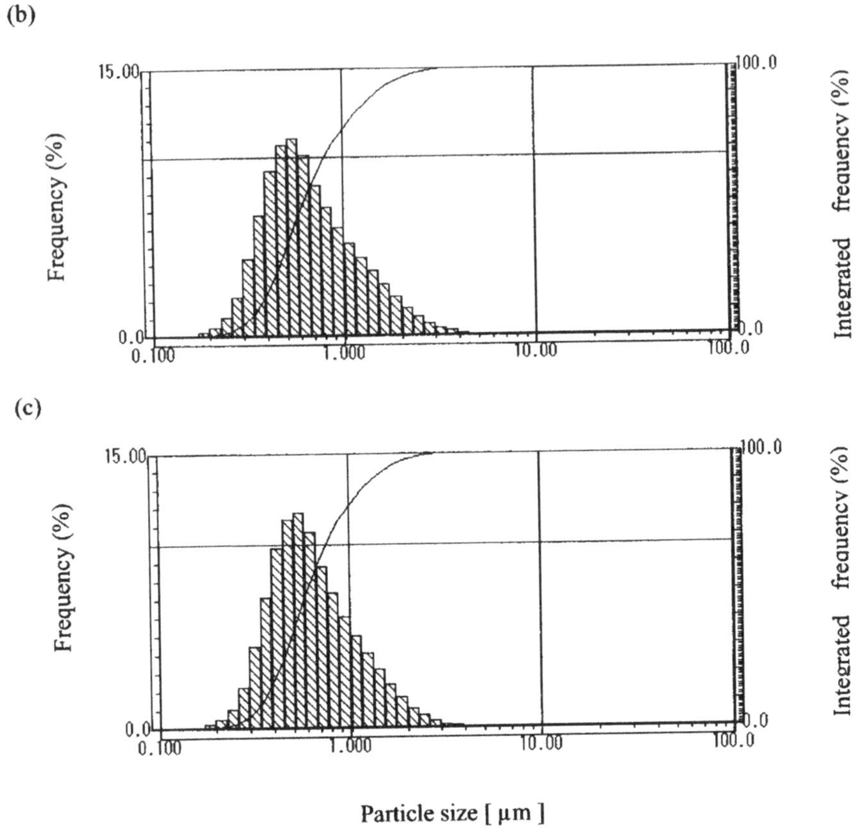

Fig. 4 Size distributions of sedimentated particles after hydrothermal treatment for 66 h without pretreatment (a)[2], 66h with pretreatment (b) and 162 h with pretreatment (c).

In case of without thermal shock pretreatment, the particle size distribution histogram had two peaks. Therefore, thermal shock pretreatment was effective to create a large surface area by introducing as many pre-cracks as possible in the sintered material. And we were able to substantially reduce the hydrothermal processing time, from 162 hours to 66 hours, compared with the case where no pretreatment was performed.

The morphology of the precipitate powders and the suspended particles, obtained after hydrothermal processing for 66 hours and 162 hours, were observed by SEM. In both cases, the precipitate particles were dispersed to such an extent that they were no different in form to the suspended particles (see Fig. 5).

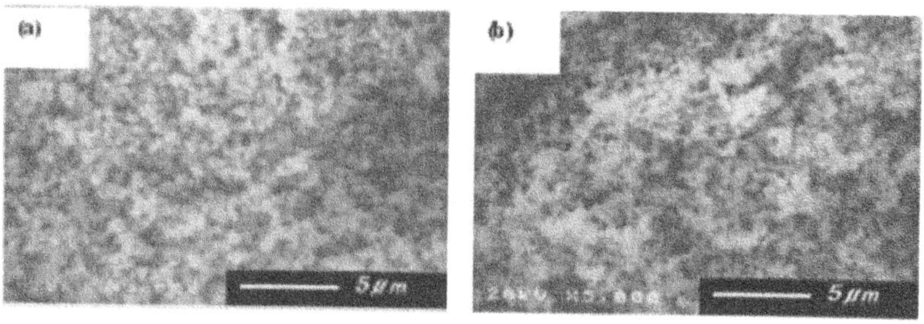

Fig. 5 SEM images of zirconia precipitate particles obtained by hydrothermal treatment. (a) after treatment for 66 h, (b) after treatment for 162 h

From the results of evaluating the powders obtained by hydrothermal processing, we decided to use precipitate powders after hydrothermal processing for both 66 hours and 162 hours in our regeneration experiments, and these powders were ground and dried using an attrition mill. Table 3 shows the changes in the characterization of the powder that took place during this grinding process.

Table 3. Changes in properties of sedimentated particles (after 66 h ,162 h treatment) by milling

Milling time (h)	0		1		2	
Treatment time (h)	66	162	66	162	66	162
BET (m^2/g)	3.5	3.5	4.4	4.3	4.9	4.7
Grain size (μm)	0.65	0.60	0.58	0.56	0.55	0.53
Monoclinic ratio (%)	90.7	91.3	90.6	90.3	90.4	90.2

The BET specific surface area increased gradually due to the grinding time, and had exceeded 4.5 m^2/g after 2 hours in both cases. Although the mean aggregate size (which indicates the dispersion state of the powder) also exhibited a downward trend, hardly any change was observed in the ratio of monoclinic crystals. Based on these results, we decided to prepare recycled powder with the attrition mill grinding time set to 2 hours.

EVALUATION OF RECYCLED Y-TZP

Using the recycled powder, sintered materials were produced by sintering at 1400°C, 1450°C and 1500°C. The ratio of monoclinic crystals at the surfaces of these sintered materials was 0vol%.

The density of the samples sintered at 1400°C was quite low — 5.84 g/cm^3 and 5.88

g/cm^3 in the samples subjected to hydrothermal processing for 66 and 162 hours respectively — but at sintering temperatures of 1450°C and above, the sample density was at least 6.0 g/cm^3. Furthermore, in the resulting recycled Y-TZP sintered materials, the strength, hardness and toughness were restored to levels comparable to those of the original bead product (see Table 4). However, a slight reduction of strength was observed in the material sintered at 1500°C, so it seems that a suitable temperature for regeneration sintering is in the region of 1450°C. And average particle size was 0.32μm in case of 66 h of hydrothermal treatment.

Table 4. Properties of recycled zirconia products, using the powder with thermal shock pretreatment, at various sintering temperatures for 2 hours.

	Y-TZP (Torayceram)	recycled product 1400°C		recycled product 1450°C		recycled product 1500°C	
Treatment time (h)		66	162	66	162	66	162
Density [g/cm^3]	6.04	5.84	5.88	6.03	6.03	6.06	6.06
Grain size [μm]	0.34±0.02	0.26±0.03	0.25±0.02	0.32±0.03	0.30±0.02	0.47±0.05	0.45±0.04
Bending Strength [MPa] • Min~Max• •	1250±55 (1190 ~ 1340)	871±114 (628 ~ 1119)	953±103 (730 ~ 1130)	975±107 (743 ~ 1143)	980±108 (731 ~ 1199)	887±84 (755 ~ 1060)	968±34 (809 ~ 938)
Hardness [GPa]	12.1±0.2	10.7±0.1	10.7±0.1	12.0±0.2	12.0±0.1	12.3±0.4	12.0±0.3
Fracture Toughness [$MPa{\cdot}m^{1/2}$]	5.5±0.2	5.5±0.2	5.4±0.2	5.4±0.1	5.4±0.2	5.5±0.2	5.5±0.2

In our previous work, we obtained fine-textured recycled Y-TZP, which showed 6.02 g/cm^3 of density, 0.32 μm of average grain size and 945 MPa of bending strength, sintered at 1450°C for 2 hours. It is clear also from this result that thermal shock pretreatment is effective in time shortening of recycle by hydrothermal processing.

Also, although the strength was generally somewhat lower than that of the original sintered material, this is thought to be because the recycled sintered material was not pelletized and contained no binder additives. In fact, the fractures were sometimes found to have started at lumps thought to be powder aggregates in strength tests on these sintered materials, especially when observing test pieces that exhibited low strength.

One of the practical difficulties involved in recycling zirconia sintered materials is that of converting zirconia sintered materials into fine powders quickly and inexpensively. According to the report which Kamiya et al.[5] leaves, the more hydrothermal corrosion advances, so that temperature is higher than 200°C. It might be effective to reduce the treatment time, but it would be desirable to carry out treatment below 200°C, when the practical and equipment problems are taken into consideration. We have shown that in order to achieve rapid disintegration of the sintered material, the process of introducing defects weakens the resistance to hydrothermal corrosion and promotes pulverization. Although this study has shown that re-usability can be made more achievable by introducing pre-cracks into unused beads, the treatment of used beads is a matter for further study. In general, grinding beads are used to disperse pigments and paints, and are employed under a wide variety of conditions. When we performed compressive failure tests on beads that had actually been used, we found that the degradation in physical properties was not all that great, and we achieved more or less the same sort of physical properties. This seems to show that used beads also require the introduction of pre-cracks by performing preprocessing in the same way as unused beads.

CONCLUSION

When Y-TZP products were subjected to a quenching process, we confirmed that it is possible to reduce the time taken to form a powder by hydrothermal processing, and that it is possible to obtain a Y-TZP recycled product from a raw material consisting of a powder with a fixed specification.

Specifically, in the state where the material has been pulverized by hydrothermal processing, the particle size distribution must have a single peak with a mean particle size of 1μm or less, and there must be no difference in yttrium concentration between the suspended particles and precipitate particles. Also, once suitable pre-cracks had been introduced, we were able to pulverize the sintered 1.5 mm diameter bead products obtained by hydrothermal processing at 200°C after hydrothermal processing for 66 hours and 162 hours.

We recovered and ground up the precipitates obtained after processing for 66 and 162 hours to yield fine powders with a mean particle size of about 0.5 μm in both cases. These powders consisted of about 90vol% monoclinic phase, caused by a transition from a tetragonal phase to a monoclinic phase during hydrothermal processing.

When the resulting fine powders were ground up, shaped and then sintered at 1400°C, 1450°C and 1500°C, it was possible to regenerate finely-textured Y-TZP products consisting almost entirely of the tetragonal phase, and no clear differences in material properties were observed between the sintered products that had been processed for 66 hours and for 162 hours.

It seems that thermal shock treatment allows the hydrothermal processing time to be reduced by 96 hours, and should lead to a reduction in processing costs.

REFERENCES

1) N.Hisamori, A. Nozue and Y. Kimura, Journal of the JSME, series A, 67, pp.844-851, (2001)
2) T. Ogata, T. Matsubara, H. Nakano, K. Urabe and T. Nishida,. Journal of the Ceramic Society of Japan, 113, (2005) to be published.
3) S. Lawson, Journal of European Ceramic Society, 15, pp.485-502, (1995)
4) J. Nakayama, “Mechanical properties of ceramics,” ed. ceramic edit committee, Gihodo, pp.69-75, (1979)
5) M. Kamiya, T. Kojima, Y. Mori, R. Sasai, and H. Itoh, “Waste Management and the Environment”, ed. D. Almoza. C. A. Brebbia, D. Sale and V. Popov, (WIT Press, Southampton, UK, pp.61-70, (2002)

Eco-Materials Processing and Design
Edited by Koji Watari and Soo-Wohn Lee

DEVELOPMENT OF NON-HEATING RECYCLING METHOD FOR USED LEAD-GLASS

R. Sasai, H. Kubo, M. Kamiya, and H. Itoh
Division of Environmental Research, EcoTopia Science Institute, Nagoya University
Furo-cho, Chikusa-ku
Nagoya, 464-8603, Japan

ABSTRACT

In the present study, we attempted to develop a non-heating recycling method for used lead-glass by a new mechanochemical technique. Pulverized lead-glass powder was sealed in a polypropyrene bottle with alumina balls, sodium ethylenediaminetetraacetate salt ($Na_2EDTA \cdot 2H_2O$), and water. After ball-milling treatment at room temperature, the slurry obtained was rinsed with distilled and deionized water, and then dried after filtration. By optimization of added amount of both $Na_2EDTA \cdot 2 H_2O$ and water, we succeeded to extraxt lead as Pb-EDTA in liquid phase from solid silicate matrix by non-heating ball-milling treatment. It was found that this separation phenomenon was controlled by the enlargement of solid-liquid interface area due to atomization of silicate matrix, and by the high stability constant of the Pb-EDTA. Under the optimal condition, elution rate of lead element was more than 99%. This high yield clarifies that Pb-O-Pb bonds in lead-glass will become loose or are broken down by ball-milling treatment, i.e., the mechanical energy such as potential or friction energy will be provided and this may be enough high to elute lead element from silicate matrix. White solid powder obtained after non-heating ball-milling treatment consisted of fine-particles of amorphous silica.

INTRODUCTION

Lead is a very important element and is utilized as various industrial materials such as battery, solder, dielectric materials and piezoelectric materials. On the other hand, it is well known that lead has a high toxicity, even when the intake is small, but if accumulated for a long term. Then, the replacement of various lead-containing materials to lead-free materials or the material recycling process of lead-containing materials has been investigated and developed by many researchers.[1-8] As a result, the use of lead-free solder becomes popular recently and the practical recycling system of lead battery has been established. However, a huge amount of lead-containing material is still used and discharged from many sectors at present. Therefore, the development of recycling and/or detoxification technology of lead containing materials is very important issues. Lead-glass is one of the very popular and useful glass materials, and is used for various applications to the protection of radioactive ray, cathode-ray tube and decorative glass. Since there is still a lot of the lead-glass consumption, a large amount of used lead-glass will be discharged. Especially, the amount of lead-glass powder from cathode-ray tube in waste TV's has increased since the law for recycling of the electric home appliance was enforced in Japan. In order to fulfill this enforcement, most of such lead-glass is landfilled after some detoxification and/or stabilization processes.

Lead atoms contained in the lead-glass are strongly fixed by encapsulation into cavities surrounded by the –O–Si–O– glass former network, and/or by directly bonded network such as –O–Si–O–Pb–O–. Thus, the Pb elution cannot be observed under normal conditions of temperature and atmospheres. Therefore, the heating temperature more than 1000°C is required to melt/evaporate and extract lead from SiO_2 glass network. The current recycling processes, partly used for lead recovery from used lead glass, need very high energy consumption, which causes a low recycling rate of lead-glass materials. Moreover, there are many kinds of lead glasses with various chemical compositions depending on their usage. Thus, it is necessary to sort these glasses by chemical composition, as far as the used lead glass was treated by the current method. This requirement will also cause lower recycling rate. In order to increase the recycling rate of the used lead glass, it is necessary to develop the novel recycling process with higher efficiency, lower energy consumption and lower price for operation and maintenance.

Ball-milling techniques, such as rotary ball-milling and planetary ball-milling, are known as one of the grinding methods for manufacturing fine powder of various inorganic raw materials. Recently, mechano-chemical method by ball-milling attracts an extensive attention for enhancing the solid-state chemical reaction or alloying with an aide of mechanical energy even at room temperature. This phenomenon occurs by the cleavage and/or formation of chemical bonding, an increase of surface area for chemical reaction, an increase of surface energy by atomizing and so on. Application of the mechano-chemical method to various waste treatments has been already reported.[9,10] In our previous study, it was found that lead atoms were selectively collected as sulfates from PZT or PLZT sintered bodies by wet ball-milling in the presence of sulfuric acid.[7,8] In this case, other elements, Zr, Ti and/or La were found to dissolve in acid solution as a water-soluble salt. However, it is rather difficult to dissolve glass components in the strong [SiO_4] network structure. Thus, we attempted to remove lead atoms from lead glass by using chelating reagent, sodium ethylenediamine *N, N', N'*-tetraacetate salt dehydrate ($Na_2EDTA \cdot 2\,H_2O$), during wet ball-milling process at room temperature. The aim of this study is the development and establishment of novel non-heating lead recycling process by combining wet ball-milling and chelate extraction.

Table 1. Chemical compositions of lead glass material

compounds*	SiO_2	PbO	BaO	K_2O	Na_2O
mass%	35.3	56.8	5.1	2.3	0.5

*) this mass ratio was calculated as oxide.

EXPERIMENTAL

Commercial lead glass for radiation shielding was used as specimen, which has average chemical composition as shown in Table 1. Sodium ethylenediamine *N, N', N'*-tetraacetate salt dehydrate ($Na_2EDTA \cdot 2\,H_2O$) purchased form DOJIN was used as chelating reagent. Coarsely ground lead glass was pretreated by the planetary ball-milling apparatus (FRITSCH P-5). Milling operation was carried out by repeating 12-times per a set, which consisted of grinding for 1.5 h at 300 rpm with an interval for 0.5 h. This pretreated lead glass, which will be abbreviated as "pre-PbG" hereafter, was dried at 110°C for 18 h under a vacuum. 2.5 g of pre-PbG powder was sealed in polypropylene bottle (50 mL), together with an appropriate amount of $Na_2EDTA \cdot 2H_2O$ and deionized/distilled water and 60 g of zirconia ball (size: ϕ5.0 mm). Ball-milling treatment of the sample was performed by a rotary ball-milling apparatus (YAMATO UB-32) at 300 rpm for 0 – 96 h. After the ball-milling, white powder formed and zirconia ball with powder were rinsed in 100 mL of water by sonication for 0.5 h, and then the white powder was collected by filtration. The white powder was dried at 110°C for 24 h under a vacuum.

Quantitative analysis of lead amount remained in lead glass was carried out by ICP-AES (Perkin-Elmer OPTIMA 3300DV) analysis of the solution obtained after dissolving glass powder by hydrofluoric acid. Lead amount in a rinsed solution was quantitatively analyzed by ICP-AES. Microstructure of lead glass particles before and after the ball-milling was observed by scanning electron microscope (SEM: JEOL JSM-6330F). Element analysis on glass particle surface was also carried out by the energy-dispersive X-ray spectroscopy (EDS: JEOL JED-2140). Specific surface area of glass powder was measured by the BET one-point method (Qantachrome MS-21). X-ray photoelectron spectroscopy was performed for determining the surface analysis of glass powder (Kratos: ESCA-3300).

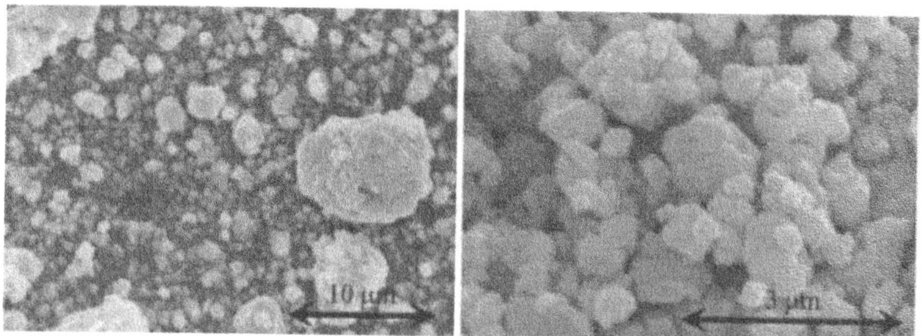

Fig. 1. SEM images of lead glass particles after pretreatment.

RESULTS AND DISCUSSION

Characterization of pre-PbG

Fig. 1 shows the SEM photographs of the pre-PbG powder. The pre-PbG powder was composed of the particles with a wide particle size distribution about 0.1 – 10 μm. Specific surface area of pre-PbG powder was 3.44 m^2/g. XPS spectra of Pb 4f, Si 2p, and Ba 3d orbits of lead glass before and after the planetary ball-milling treatment are shown in Fig. 2. Peak of Pb 4f orbit shifted to lower energy side after the planetary ball-milling process (cf. Fig. 2(a)). It is well known that there are two species of lead in lead glass, modification (139.0 eV) and formation (138.2 eV) types for glass network.[11] The modification type was greater than the formation type in the present lead glass. However, the amount of the formation type increased after planetary ball-milling. The difference of this abundance ratio between modification and formation types causes the peak energy shift, which indicates that the formation type might increase by the planetary ball-milling process. Peak energy shift of Si 2p orbit spectrum to the high energy side was observed after the planetary ball-milling process (cf. Fig. 2(b)). As a result of spectrum deconvolution, it was found that the observed peak consisted of three bands from $[SiO_4]^{4-}$, $[Si_4O_{12}]^{8-}$, Si–O–Pb species, respectively. Peak shift could be caused by the change of the abundance ratio among this three species. After the planetary ball-milling, the abundance amount of the Si–O–Pb species from the formation type decreased. This result indicates that lead atom of formation type in lead glass decreased by planetary ball-milling, but this result would contradict the result from XPS spectra of Pb 4f orbit. Here, peak energy of PbO (138.3 eV), which will be formed by the cleavage of Si–O–Pb network, is almost the same as the peak energy of lead atoms of modification type (138.2 eV). By considering this information, the increase of modification-type lead atoms observed in XPS spectra of Pb 4f would arise from the formation of PbO. Therefore, the planetary ball-milling could induce the isolation of PbO from Si–O–Pb network by the applied mechanical energy such as friction and/or collision energy.

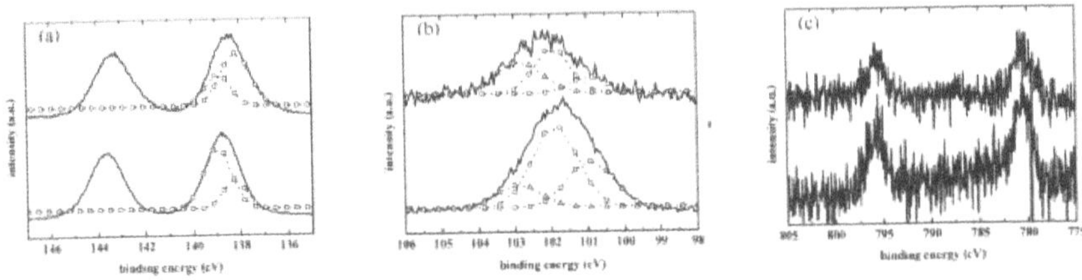

Fig. 2. XPS spectra of (a) Pb 4f, (b) Si 2p and (c) Ba 3d orbits in lead glass before (upper) and after (bottom) the pretreatment. Symbols in (a) mean the modification type (squares) and formation type (circles) of lead atom in lead glass. Symbols in (b) mean the species of $[SiO_4]^{4-}$ (triangles), $[Si_4O_{12}]^{8-}$ (circles) and Si-O-Pb (squares).

XPS spectra of Ba 3d before and after the planetary ball-milling did not change (cf. Fig. 2(c)). It is well known that Ba ions play a role as the modifier of glass network. This result also supports that the planetary ball-milling process could result in the bond formation, but the cleavage of glass network.

Extraction of Pb from pre-PbG by wet ball-milling with $Na_2EDTA \cdot 2H_2O$

The specific surface area of glass powder obtained by the ball-milling treatment for 45 h with various Na_2EDTA contents are plotted against the water/solid fraction, as shown in Fig. 3, where the solubility of $Na_2EDTA \cdot 2H_2O$ for water were as 11.1g/100mL. Specific surface area increased with an increase in water content, and then saturated at more than ca. 0.5 ($Na_2EDTA \cdot 2H_2O$ = 2.5 g) or ca. 0.7 mL/g ($Na_2EDTA \cdot 2H_2O$ = 3.0 g), but no saturation behavior was observed in the case of ball-milling treatment with 3.5 g of Na_2EDTA. The increase of specific surface area is due to the grain refinement of lead glass particle. Thus, this saturation behavior indicates that there are some limits for grain refinement by the ball-milling with 2.5 and 3.0 g of $Na_2EDTA \cdot 2H_2O$. Moreover, the difference of the grain-refinement behavior was observed, i.e., the less than ca. 0.65 mL/g of the water/solid fraction became, the more amount of water was

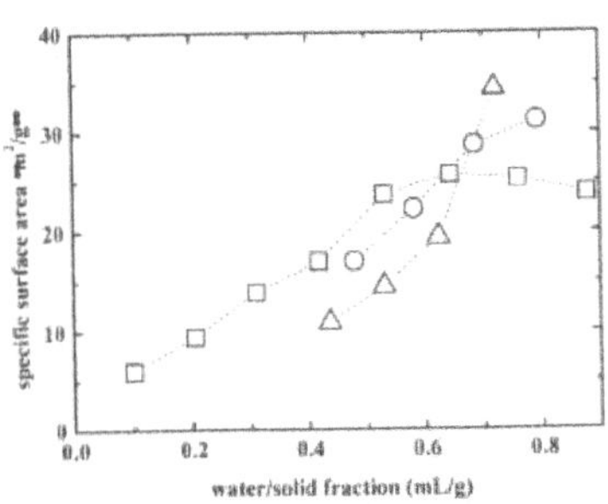

Fig. 3. Dependence of specific surface area on the water/solid fraction. $Na_2EDTA \cdot 2H_2O$ content is 2.5 (squares), 3.0 (circles) and 3.5 (triangles) g.

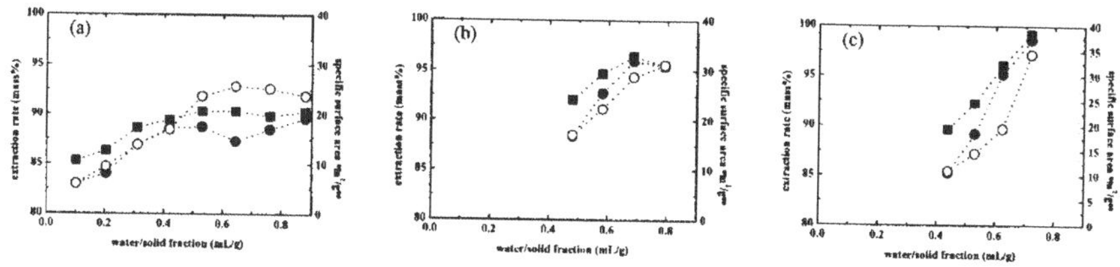

Fig. 4. Dependence of the extraction rate (filled symbols and left axis) of Pb (squares) and Ba (circles) and specific surface area (open symbols and right axis) on the water/solid fraction after ball-milling for 45 h. $Na_2EDTA \cdot 2\,H_2O$ are 2.5 g (a), 3.0 g (b) and 3.5 (c) g.

required for obtaining the finer particle by wet ball-milling with an increase in the added $Na_2EDTA \cdot 2\,H_2O$. However, the amount of water required for refining lead glass powder decreased with an increase in the added $Na_2EDTA \cdot 2\,H_2O$ more than ca. 0.65 mL/g because of the saturation behavior of specific surface area. It is difficult to explain this difference of grain refinement behavior by only water/solid fraction, especially, less than ca. 0.65 mL/g.

Fig. 4 shows the dependence of extraction rate of Pb and Ba from lead glass on the water/solid fraction. It is found that the wet ball-milling treatment in the presence of $Na_2EDTA \cdot 2\,H_2O$ can remove Pb and Ba atoms from lead glass particles. The extraction behavior exhibits a similar tendency as that of the specific surface area. This result indicates that the extraction of Pb and Ba atoms from lead glass powder is caused by the formation of EDTA chelate complexes with Pb and/or Ba ions exposed on particle surface by ball-milling because of higher stability constant of Pb-/Ba-EDTA complex than Na-EDTA complex. Extraction rate of both Pb and Ba atoms was constant at ca. 90 mass%, which is more than ca. 0.4 mL/g of the water/solid fraction, when added amount of $Na_2EDTA \cdot 2\,H_2O$ is 2.5 g. This discrepancy between extraction rate and specific surface area can be explained by the shortage of added $Na_2EDTA \cdot 2\,H_2O$, i.e., added amount of $Na_2EDTA \cdot 2\,H_2O$ (6.71 mmol) was lower than the summation of Pb and Ba in lead glass powder (7.26 mmol). This consideration is supported by the fact that the summation of Pb and Ba remained in treated lead glass powder is 0.51 mmol (Pb: 0.5 mmol and Ba: 0.01 mmol). The saturation behavior of Pb- and Ba-extraction rate was also observed in the case of adding 3.0 g of Na_2EDTA, although the added amount of $Na_2EDTA \cdot 2\,H_2O$ was enough to extract Pb and Ba atoms from lead glass powder. This saturation behavior is corresponding to that of specific surface area, which suggests that no pulverization of glass particles is possible, i.e., the probability of Pb and Ba atoms found out on particle surface will become lower. These results indicate that the extraction of Pb and/or Ba by EDTA chelation will occur on the only glass particle surface. It will be possible, however, to remove the all Pb and/or Ba contained in the present lead glass by ball-milling, if very fine particles are prepared by ball-milling process. These conclusions are also supported by Fig. 4(c), where the tendency of the Pb- and Ba-extraction rate is corresponding to that of specific surface area that increases with an increase in water/solid fraction.

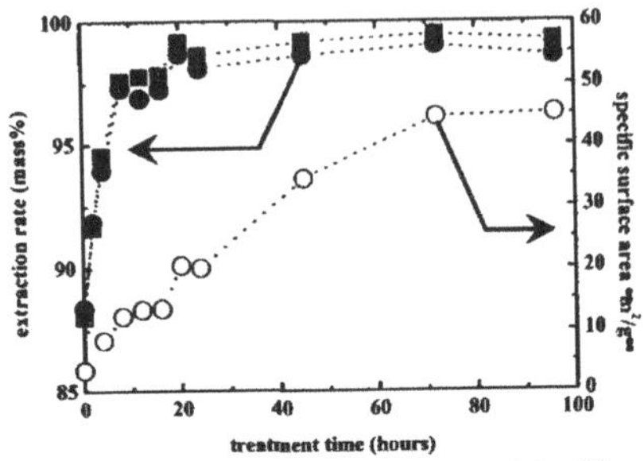

Fig. 5. Treatment time dependence of the Pb- (squires) and Ba- (circles) extraction rate (filled symbols: left axis) and specific surface area (open symbols: right axis). $Na_2EDTA \cdot 2\,H_2O$: 3.5 g. water: 4.0 mL.

Fig. 5 shows the treatment time dependence of Pb- and Ba-extraction rate and specific surface area after wet ball-milling with 3.5 g of $Na_2EDTA \cdot 2\,H_2O$ and 4.0 mL of water. About 88 mass% of Pb and Ba is extracted at the sonication step of pre-PbG, which means a sample powder without wet ball-milling. This result shows that ca. 88 mass% of Pb and Ba contained in lead glass is already exposed as ions loosely bounded to the surface of glass particles after the planetary ball-milling treatment and they are extracted by the chelate formation with EDTA. The specific surface area gradually increased with an increase in the treatment time and became almost constant at more than 72 h. On the other hand, the extraction rate of Pb and Ba rapidly increased up to 8 h, then gradually increased with an increase in the treatment time and became almost constant at more than 72 h. It was elucidated from these results that the optimal treatment time was 72 h under the present conditions ($Na_2EDTA \cdot 2\,H_2O$: 3.5 g, water: 4.0 mL), when we succeeded in removing 99.5 mass% of Pb ant 99.0 mass% of Ba from lead glass by ball-milling treatment without heat treatment.

In Fig. 6, the SEM images of lead glass particles after wet ball-milling treatment under optimal condition for various treatment times. The pre-PbG particles had relatively wide grain size distribution (cf. Fig. 6(a)). From SEM images of lead glass particles for 4 and 8 h, it is apparent that the large particles observed in Fig. 6(a) are preferentially pulverized by wet ball-milling (cf. Figs. 6(b) and (c)). The Pb and Ba atoms inside of the coarse particles could be exposed to the outer surface of the pulverized particles and were extracted by the chelate formation with EDTA ligands. This preferential pulverization of coarse particles would promote the rapid increase in Pb- and Ba-extraction rate up to 8 h. After 24 h ball-milling, the particle size gradually decreased, which resulted in a small exposure of Pb and Ba atoms on particle surface, and thus the extraction rate increased slightly. Finally, the extraction rate was saturated because the exposure of fresh surface on lead glass particle finished even by wet ball-milling. This is supported also by the saturation of specific surface area at more than 72 h.

CONCLUSIONS

We succeeded originally in the removal of Pb and Ba atoms from used lead glass by mechano-chemical method using a wet ball-milling without heat treatment. Optimization of treatment conditions clarified that ca. 99.5 mass% of Pb and ca. 99.0 mass% of Ba were recovered as Pb-EDTA and Ba-EDTA complexes, respectively. Unlike the current recycling

Fig. 6. SEM images of lead glass particles after wet ball-milling ($Na_2EDTA \cdot 2H_2O$: 3.5 g, water: 4.0 mL) for (a) 0, (b) 4, (c) 8, (d) 24, (e) 45, and (f) 72 hours.

process for the used lead glass, the present method has a lot of advantages: (1) the energy consumption is very low, because the treatment is carried out at room temperature and (2) the existing ball-milling apparatus for fine-powder production can be applicable to this recovery system of lead from used lead glass. Furthermore, the present process will be effective for recycling for not only lead glass, but also other glass materials containing toxic or precious metals.

REFERENCES

[1]M. Kitajima, and T. Shono, "Reliablity Study of New SnZnAl Lead-Free Solders Used in CSP Packages," *Microelectorn. Relaiab.*, **45**, 1208-14 (2005).

[2]X. Y. Zhou, H. S. Gu, and W. Y. Li, "Effect of Doped Mn on Piezoelectrc Properties of (Na0.5Bi0.5)(0.92)Ba0.08Ti3 Lead-Free Ceramics," *J. Cnet. South Univ. T.*, **12**, 266-8 (2005).

[3]X. X. Wang, S. H. Choy, X. G. Tang, and H. L. W. Chan, "Dielectric Behavior and Microstructure of (Bi1/2Na1/2)TiO3-(Bi1/2K1/2)TiO3-BaTiO3 Lead-Free Piezoelectric Ceramics," *J. Appl. Phys.*, **97**, 104101 (2005).

[4]A. J. Saterlay, S. J. Wilkins, and G. Compton, "Towards Greener Disposal of Waste Cathode Ray Tubes via Ultrasonically Enhanced Lead Leaching," *Geen Chem.*, 3, 149-55 (2001).

[5]H. Miyoshi, D. Chen, and T. Akai, "A Novel Process Utilizing Subcritical Water to Recycle Soda-Lime-Silicate Glass," *J. Non-Cryst. Solids*, **337**, 280-2 (2004).

[6]H. Miyoshi, D. Chen, and T. Akai, "A Novel Process Utilizing Subcritical Water to Remove Lead from Wasted Lead Silicate Glass," **33**, 956-7 (2004).

[7]M. Kamiya, R. Sasai, and H. Itoh, "Recovery of Lead in Lead Zirconate Titanate Ceramics by Wet Ball Milling with Acidic Solution," *J. Ceram. Soc. Jpn.*, **111**, 806-10 (2003).

[8]M. Kamiya, R. Sasai, and H. Itoh, "Recovery of Lead from PLZT Ceramics Using Wet Ball-Mill Method," *J. Ceram. Soc. Jpn.*, **111**, 897-901 (2003).

[9]V. Birke, J. Mattik, and D. Runne, "Mechanochemical Reductive Dehalogenation of Hazardous Polyhalogenated Contaminants," *J. Mater. Sci.*, **39**, 5111-6 (2004).

[10]R. B. Rao and B. P. Singh, "Effective of Oxidative Degradation of Coal on Ball Milling of Basic Slag," *Metals Mater. Proc.*, **9**, 45-8 (1997).

[11]O. M. Kanunnikova, F. Z. Gilmutdinov, and A. A. Shakov, "Interaction of Lead Silicate Glasses with Hydrogen under Heating," *Int. J. Hydrogen Energ.*, **27**, 783-91 (2002).

Eco-Materials Processing and Design
Edited by Koji Watari and Soo-Wohn Lee

THE EFFECT OF CRYSTAL PHASE FORMATION ON LEACHABILITY OF Pb FROM GLASS CERAMICS PREPARED FROM INDUSTRIAL ZINC WASTE

Bussaraporn Patarachao and Sirithan Jiemsirilers 1
Research Unit of Advanced Ceramics, Department of Materials science, Faculty of Science. Chulalongkorn University, Bangkok 10330 Thailand 1

Parjaree Thavorniti and Sitthisuntorn Supothina 2
National Metal and Materials Technology Center, 114 Thailand Science Park, Paholyothin Rd., Klong Luang, Pathumthani 12120 Thailand 2

ABSTRACT

The correlation between the crystalline phase present in glass-ceramics and the leachability of lead (Pb) was investigated. In this study, glass-ceramics were prepared by the crystallization of glasses obtained by melting raw materials containing 30% of zinc waste from a hydrometallurgical plant mixed with other oxides. The crystalline phases present in the glass-ceramics were identified by x-ray diffractometry (XRD), while the crystal morphologies were investigated by scanning electron microscopy (SEM) and the composition of crystals and residual glasses were confirmed by energy-dispersive x-ray diffractometry (EDS). The leachability of lead (Pb) was examined following the standard procedure of the United States Environmental Protection Agency Toxic Characterization Leaching Procedure (TCLP). The leached out concentration of Pb was then quantified by using atomic absorption spectroscopy (AAS). It was found that the glass-ceramics containing wollastonite-ferroan phases were more stable towards acid leaching, whereas those containing pyroxene phase were less stable.

INTRODUCTION

In recent years, recycling of industrial wastes such as municipal solid waste, incinerator fly ash, blast furnace slag, etc. have been studied by many scientists around the world.[1-5] Glass and glass-ceramic technologies are considered to be the most appropriate technologies to immobilize heavy metals from these wastes in the stable glass matrix with a consistent reduction of volume and the destruction of organic matter associated with the wastes during glass processing.[6,7] In Thailand, the company that produces zinc ingot is located at Tak province. Each year zinc refinery process produces large amounts of waste, which contains significant percentage of Lead (Pb). The possibility of transformation of this zinc waste into more valuable products such as glass-ceramic material has been studied.[8] In this study the effect of crystal phase formation on leached out concentration of Pb from glass-ceramics is the main concerned. The leachability of Pb was examined to determine the quality of obtained glass-ceramics.

EXPERIMENT PROCEDURE

Characterization of zinc waste

Initially, chemical compositions of zinc waste used in this study was analyzed by using x-ray fluorescence (XRF) since the compositions of zinc waste generated from different plants can be varied depending on the type and the source of zinc minerals, as well as the materials adding during zinc recovery process.[9,10]

Preparation of glass

Glass compositions were designed by using Mixsoft Computer software[11] based on eight oxides which were SiO_2, B_2O_3, Na_2O, CaO, MgO, ZnO, Al_2O_3 and Fe_2O_3. Glass compositions are shown in Table I.

Table I Glass compositions in weight percent of oxides

	Composition (%wt)							
	SiO_2	B_2O_3	Na_2O	CaO	MgO	ZnO	Al_2O_3	Fe_2O_3
G1	45.00	15.00	15.00	20.00	0.00	1.00	1.00	3.00
G2	55.00	5.00	13.00	20.00	0.00	1.00	1.00	5.00
G3	55.00	5.00	15.00	18.00	0.00	1.00	1.00	5.00
G4	45.00	15.00	13.00	20.00	0.00	3.00	1.00	3.00
G5	45.00	15.00	11.00	20.00	0.00	1.00	5.00	3.00
G6	63.00	5.00	15.00	10.00	0.00	3.00	1.00	3.00
G7	64.00	5.00	10.00	10.00	0.00	3.00	5.00	3.00
G8	45.00	15.00	5.00	18.00	10.00	3.00	1.00	3.00
G9	45.00	15.00	10.00	10.00	9.00	1.00	5.00	5.00
G10	45.00	15.00	11.00	10.00	10.00	3.00	1.00	5.00
G11	45.00	15.00	11.00	10.00	10.00	1.00	5.00	3.00
G12	56.00	5.00	5.00	15.00	10.00	3.00	1.00	5.00
G13	55.00	5.00	15.00	10.00	8.00	1.00	1.00	5.00
G14	55.00	5.00	15.00	10.00	10.00	1.00	1.00	3.00

In each batch thirty percent of zinc waste was weighted and mixed with other raw materials to produce suitable glass compositions. The mixture was melted at 1450°C for 1 hour to form glasses.

Preparation of glass-ceramics

Preparation of glass-ceramic was performed by heat-treatment process including two main steps. Firstly, the samples were heat-treated up to 750°C and hold at this temperature for 1 hour, and then further heat-treated up to 850°C and hold at this temperature for 2 hours.

Characterization of glass-ceramics
Crystal phase analysis

After heat-treatment, crystalline phases of glass-ceramics were identified by using an x-ray diffractometer (XRD). The size and morphology of the crystals were observed by using a scanning electron microscope (SEM). The chemical compositions of crystal phase and glassy phase contained in glass-ceramics were examined by using an energy-dispersive x-ray spectrometer (EDS).

Chemical durability

Since the zinc waste contained significant amount of lead (Pb), therefore it was very important to examine the leachability characteristic of Pb in glass-ceramics matrix. Noted that the regular standard of heavy metal leached out concentration is different in difference countries.[5-7] In this study the American EPA's toxic characteristic leaching procedure (TCLP) method 1311, which is the acid leaching method, was used to examine the chemical stability of glass-ceramics in terms of leaching of heavy metals ions.[12] The leached out concentration of Pb was then examined by using an atomic absorption spectrometer (AAS).

RESULTS AND DISCUSSION

Chemical compositions of zinc waste

The chemical composition of zinc waste was analyzed quantitatively by using XRF. The compositions in weight percent of oxides are shown in Table II.

Table II Chemical compositions of zinc waste analyzed by XRF

Oxides	%wt
SiO_2	22.45
Al_2O_3	3.96
Fe_2O_3	7.32
CaO	16.43
MgO	0.29
K_2O	0.63
ZnO	3.09
CuO	0.07
PbO	1.07
TiO_2	0.10
MnO	0.10
P_2O_5	0.21
ZrO_2	0.01
SO_3	30.64
L.O.I	13.94

Zinc waste was composed of three main oxides, which were SiO_2, CaO and SO_3, with respect to the zinc refinery process.

Crystal phase formation of glass-ceramics

Five different types of crystalline phase have been identified by XRD as follows: wollastonite-ferroan, anorthite, quartz, cristobalite and pyroxene. Single phase of wollastonite-ferroan was found in G1, G2 and G3. Single phase of pyroxene was found in G8 – G14. Mixed phases of wollastonite-ferroan and anorthite were found in G4 and G 5. Mix phases of wollastonite-ferroan, quartz and cristobalite were found in G6 and G7. The XRD patterns of all five crystalline phases in the selected samples are shown in Fig. 1.

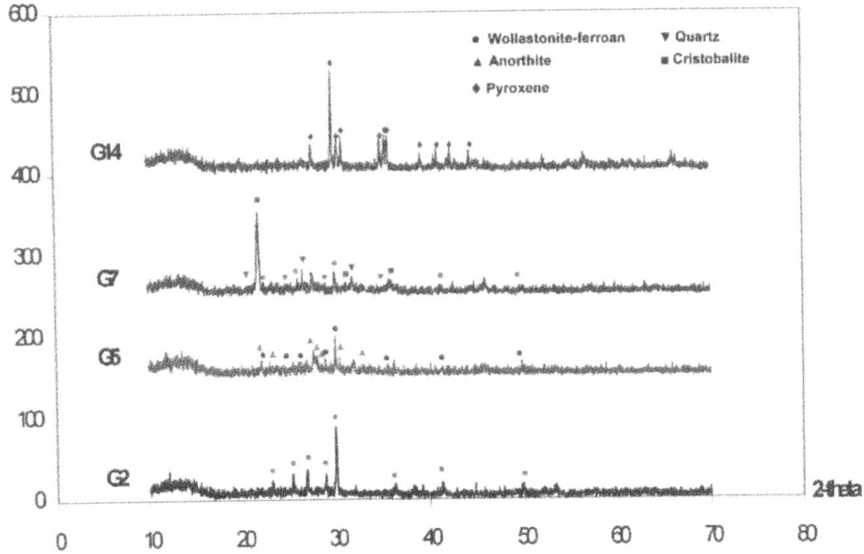

Fig. 1 XRD patterns of glass-ceramics G2, G5, G7 and G14

As described above that single pyroxene phase was found in G8-G14, these glass-ceramics contained 8-10wt%MgO, whereas other phases were found in non-adding MgO glass-ceramics. This is believed to be due to the appropriate of chemical composition of the parent glass with the suitable time and temperature.

The SEM micrograph of G2, G5, G7 and G14 are shown in Fig. 2 (a), (b), (c) and (d), respectively. Acicular crystals dispersed in the glassy matrix were observed in G2. The EDS analysis showed that acicular crystals contain mainly Si, Ca and confirming wollastonite-ferroan ($Ca_{2.87}FeO_{0.13}(SiO_3)_3$) whereas glassy matrix enriched in Si, Al, Na and Fe with small amount of Ca.

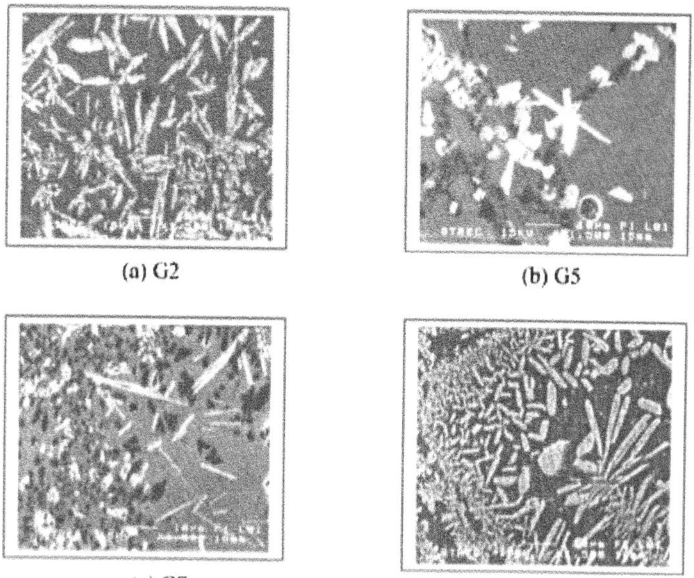

(a) G2 (b) G5

(c) G7 (d) G14

Fig.2 SEM micrograph of the crystalline phases

In fig 2 (b) (G5), two distinct crystal morphologies crystals were observed. The acicular crystals contain mainly Si, Ca and Fe confirming wollastonite-ferroan, the dark dendritic crystals contain mainly Si, Al and Ca confirming anorthite ($Ca(Al_2SiO_8)$). The acicular crystals enriched in Si, Ca and Fe of wollastonite-ferroan.were also observed in G7.

Dendritic crystals were observed in the SEM micrograph of G14. The EDS analysis showed that the dendritic crystals contain mainly Si, Mg, Ca and Fe confirming pyroxene $(Mg_{0.937}Fe_{0.063})(Ca_{0.751}Na_{0.249}Fe_{0.018})(Si_2O_6)$.

Chemical durability

The effect of crystalline phase on leached out concentration of Pb was studied. The leached out concentration of Pb from glass-ceramic matrix are shown in Table III. The concentration of Pb in the leachate of glass-ceramics was compared with the USA regulatory limit (5 ppm). Note that the TCLP test was conducted on the pulverized specimens.

Table III Leached out concentration of Pb in glass-ceramics

	Crystalline phase	Concentration (ppm)
G1	W	2
G2	W	2
G3	W	2
G4	W/A	2
G5	W/A	2
G6	W/C/Q	2
G7	W/C/Q	2
G8	P	22
G9	P	32
G10	P	42
G11	P	26
G12	P	23
G13	P	20
G14	P	18

Note: W is Wollastonite-ferroan, A is Anorthite, C is Cristobalite phase, Q is Quartz

According to Table III, it is very interesting that the leached out concentrations of Pb in glass-ceramics contained pyroxene phase were so much higher than those of glass-ceramics contained wollastonite-ferroan phase. None of glass-ceramics contained pyroxene phase gave the leached out concentration of Pb lower than 5 ppm, whereas glass-ceramics contained wollastonite-ferroan phase gave the leached out concentration of Pb lower than the limit. Not only glass-ceramics contained single wollastonite-ferroan phase, glass-ceramics contained multiple phase of wollastonite-ferroan + anorthite (W/A) and wollastonite-ferroan + cristobalite + quartz (W/C/Q) also gave the leached out concentration of Pb lower than 5 ppm. The leached out concentrations of Pb from glass-ceramics contained pyroxene phase are varied from 18 – 42 ppm. Therefore it is evident that crystal phase formation in glass-ceramics strongly effected on the leachability of Pb.

In this study, it is clear that those glass-ceramics contained pyroxene phase are unstable toward the acid leaching which means they are poor in chemical durability whereas glass-ceramics contained both single and multiple of wollastonite-ferroan phase are much more stable toward the acid leaching.

CONCLUSION

Glass-ceramics obtained in this study are composed of many types of crystalline which are wollastonite-ferroan, cristobalite, quartz, anorthite and pyroxene depending on glass compositions. The crystal phase present in the glass-ceramic strongly effects on a leachability of Pb. The glass-ceramics contained pyroxene phase are very poor in chemical durability and unstable toward acid leaching whereas glass-ceramics contained single and multiple phase of wollastonite-ferroan are stable.

ACKNOWLEDGEMENT

We are very grateful to the National Metal and Materials Technology Center (MTEC), the National Science and Technology Development Agency (NSTDA), Thailand for a financial support of this project. We also would like to acknowledge Padaeng Industry Public Co. Ltd. for supporting zinc waste as a raw material used in this study. Miss Patarachao would like to acknowledge the Thailand Graduate Institute of Science and Technology (TGIST) for financial support during an M.S. study.

REFERENCES

[1]Alexander Karamanov, Mario Pelino and Alessandro Hreglich, "Sintered glass-ceramics from municipal solid waste-incinerator fly ashes-part I: the influence of the heating rate of the heating rate on the sinter-crystallization," *Journal of the European Ceramic ociety* **23,** 827-832 (2003)

[2]Alexander Karamanov, Mario Pelino, Milena Salvo and Ilidko Metekovits, "Sintered glass-ceramics from incinerator fly ashes. Part II. The influence of the particle size and heat-treatment on the properties," *Journal of the European Ceramic Society* **23,** 1609-1615 (2003).

[3]T.W. Cheng, T.H. Ueng, Y.S. Chen and J.P Chiu, "Production of glass-ceramic from incinerator fly ash," *Ceramics International* **28,** 779-783 (2002)

[4]Young Jun Park and Jong Heo, "Conversion to glass-ceramics from glasses made by MSW incinerator fly ash for recycling," *Ceramics International* **28,** 689-694 (2002)

[5]Michihiro Miyake, Chikashi Tamura and Motohide Matsuda, "Resource Recovery of Waste Incineration fly Ash: Synthesis of Zeolites A and P," Journal of the American Ceramic Society 85, 1873 (2002)

[6]P. Pisciella, S. Crisucci, A. Karamanov, M. Pelino, "Chemical durability of glasses obtained by vitrification of industrial wastes," *Waste Management* **21**, 1-9 (2001)

[7]Mazimina Romero and Jesus Ma. Rincon, "Surface and bulk crystallization of glass-ceramic in the Na_2O-CaO-ZnO-PbO-Fe_2O_3-Al_2O_3-SiO_2 system derived from a goethite waste," *Journal of the American Ceramic Society* **82,**1313-1317 (1999)

[8]S. Jiemsirilers, B. Patarachao, P. Thavornniti, and S. Supothina, "Properties of Glass-ceramics synthesized from hydrometallurgical Waste," *Melt Chemistry, Relaxation and Solidification Kinetics of Glasses, Ceramic Transactions* **170**, 237-246 (2005)

[9]M. Pelino, "Recycling of jarosite waste in the production of glass and glass-ceramic materials," *Interceram* **47**, 22-26 (1998)

[10]M. Pelino, "Recycling of zinc hydrometallurgical wastes in glass and glass-ceramic materials," *Waste Management* **20**, 561-568 (2000)

[11]G.F. Piepel, "*Mixsoft-software for the design and analysis of mixture and other constrained region experiments,*" User's guide version 2.31, The United States of America (2000)

[12]Method 1311 Toxic characteristic leaching procedure (online), Available from: http://www.epa.gov/epaoswer/hazwaste/test/pdfs/1311, July (1992)

Eco-Materials Processing and Design
Edited by Koji Watari and Soo-Wohn Lee

HIGH PERCENTAGES OF HYDROMETALLURGICAL ZINC WASTE LOADING IN UNGLAZED TILE BODY

Laksana kreethawate and Sirithan Jiemsirilers
Research Unit of Advanced ceramics, Department of Materials science, Faculty of Science, Chulalongkorn University, Bangkok 10330 Thailand;

Parjaree Thavorniti and Sitthisunthon Supothina
National Metal and Materials Technology Center, 114 Thailand Science Park, Pathumthani 12120 Thailand

ABSTRACT

The possibility of using hydrometallurgical zinc waste as raw material in fabricating ceramic unglazed tiles was investigated in order to immobilize hazardous elements present in this waste. The compositions were prepared by mixing hydrometallurgical zinc waste from 10% to 90% with traditional tile body. The specimens were made by dry-pressing and then fired at temperatures range from 1100-1150°C. The properties of fired specimens such as water absorption, linear shrinkage, bulk density, three-point bending flexural strength and chemical resistance were determined and the leaching test was also done. It was observed that water absorption obtained to zero, high bending strength, good chemical resistance and low leaching of toxic. The introduction of up to 70 % of zinc waste into the unglazed tile body showed acceptable properties for this application.

INTRODUCTION

Zinc hydrometallurgical process at Tak plant, Thailand generated large amount of zinc waste. Due to this waste contains hazardous heavy metals; it produces many economical and environmental problems. Recycle technology is a well known technique to solve these problems. Nowadays, literatures have reported on the utilization of zinc waste as a major raw material for some products such as building materials, pigments, refractories and glass-ceramics.[1-3] Our previous work, zinc waste was converted to glass via vitrification process and transform to glass ceramics by heat treatment[4]. The results of them show mechanical properties and chemical durability can be improved by crystallization process. Moreover zinc waste also contains many compounds that are possible for using in tile production. The aim of this study is to reuse high loading of this waste as materials for formulating unglazed tile body. This production is lower cost than vitrification process since it consumes lower power than vitrification technique. Further more, no body stain is needed in this experiment. Various colors of these unglazed tiles came from different percent of zinc waste. The study is also concerned about the effect of compositions on properties of unglazed tile.

EXPERIMENTAL PROCEDURE

Raw Material Characterization

The chemical composition of zinc waste was examined by using X-ray fluorescence spectroscopy (XRF, Phillips Pw-2400) and phase analysis of this waste was determined by using X-ray diffraction (XRD, Bruker D8 Advance).

Tile Preparation

Batch composition of the tile bodies were formulated by using a traditional tile body mixed with zinc waste in different proportions according to Table I. Each batch of tile composition was prepared by wet milling in high speed pot mill for various milling times to obtain a residue of 3 wt% on a sieve of #230 mesh (63 μm). After that the slurry was dried at 110°C and ground into powder. The moisture content of the dried powder was adjusted to have a moisture content of 5 %, aged for 24 h in a closed container and then sieved through a sieve #35 mesh (500 μm). To fabricate the tile, the powder was pressed by uniaxial pressing machine of 270 Kg/cm^2 pressure. After drying at 110°C, the specimens were fired at 1100, 1125 and 1150°C in an electric furnace at heating rate of 5°C/min and with soaking time of 20 min.

TableI. Batch composition of specimens

Specimen	% Zinc waste	% Tradition tile body
T1	10	90
T2	20	80
T3	30	70
T4	40	60
T5	50	50
T6	60	40
T7	70	30
T8	80	20
T9	90	10

Tile Characterizing

Fired specimens for general properties such as surface, shape and color were investigated by visual observation. Linear shrinkage was determined by measuring the lengths of specimens before and after firing. Density, water absorption and apparent porosity were measured by Archimedes method according to the procedure described in ASTM C373.[5] The specimens fired at various temperature were selected based on their water absorption follows in the range of standard unglazed tile for further experimentation such as strength, phase analysis, chemical resistance, and Pb leaching. The strength of the selected specimens were measured by using a 3-point bending method with a span of 45 mm on an universal testing machine using a rectangular bar of dimension 50mm x 10mm x 5mm. The chemical resistance was tested according to ASTM C650[6]. The release of the hazardous element (Pb) was determined by following toxicity characteristic leaching procedure (TCLP 1311) method[7]. Finally, the phases present in the specimens fired at 1125°C were determined by using an XRD.

RESULTS AND DISCUSSION

Properties of Zinc Waste

According to the XRF result shown in Table 2, this waste composed of a SiO_2 as a major component. Minor oxides such as CaO, ZnO, Fe_2O_3, Al_2O_3 and SO_3 are also present. In term of mineral composition, this waste composed of quartz (SiO_2), cristobalite (SiO_2) anhydrite ($CaSO_4$) and Magnetite (Fe_2O_3).

TableII. Chemical composition of zinc waste (wt%)

Oxides	SiO_2	Al_2O_3	Na_2O	MgO	K_2O	CaO	Fe_2O_3
Wt%	57.09	4.09	0.01	1.19	0.83	10.25	7.545
Oxides	ZnO	SO_3	PbO	TiO_2	P_2O_5	MnO	CuO
Wt%	9.53	4.81	1.23	0.19	0.10	0.54	0.17

Tile Properties

Color and Appearance

The color of the fired specimen was found to be dependent on both composition and firing temperature. At any selected firing temperature, the color shades of the tile specimens changed from brown to dark brown. The change of color depended on the amount of zinc waste - the higher percentage of the zinc waste, the darker the color of the tile. That is, the darkness increased from the specimen T1 to T9. At the same composition, the color shade of the specimens depended on the firing temperature. Increased firing temperature resulted to darker specimens.

There was no shape distortion observed in all the specimens at firing temperatures up to 1125°C. However, after firing at 1150°C, the bloating were observed in specimens of T4, T5, T6, T7, T8 and T9, all of which possessed high content (≥ 40%) of zinc waste.

Physical Properties

The effect of firing temperature on a linear shrinkage, bulk density and water absorption of the tile specimens are shown in Figs. 1, 2 and 3, respectively. Linear shrinkage and bulk density of the specimens T1, T2, T3, T4 and T9 increased with increasing firing temperature. For the specimens T5, T6 and T7, a linear shrinkage and density decreased with increasing temperature. The shrinkage and density of the specimen T8 increased with temperature up to 1125°C, and then slightly decreased afterward. Because of overfiring the shrinkage and bulk density of specimens at high temperature decreased.

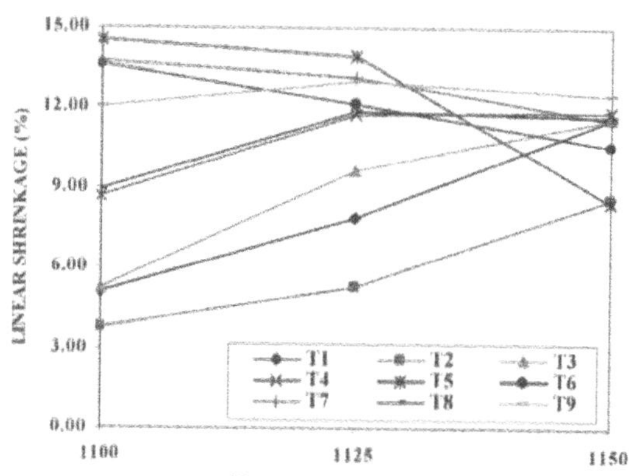

Fig1. Linear shrinkage of specimens

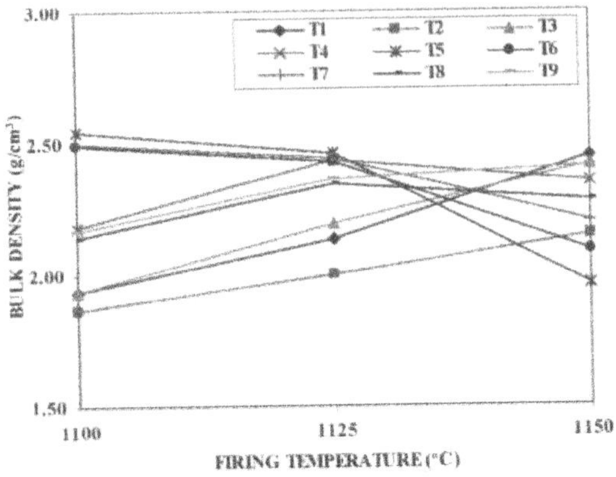

Fig2. Bulk density of specimens

Water absorption of specimens T1, T2, T3, T9 (1100 – 1150 °C), T4 and T8 (1100 – 1125 °C) decreased as firing temperatures increased. For specimens T4 and T8 after 1125 °C their water absorption slightly increased. Specimens T5, T6 and T7 obtained zero absorption when fired at as lower as 1100 °C, increasing firing temperatures caused water absorption to increase due to overfiring. Higher percentages of zinc waste in specimens yield lower water absorption at lower firing temperature. Water absorption of T5, T6 and T7 were obtained to zero at 1100°C. While, water absorption of T1, T2, T3 and T4 were obtained to zero at higher temperatures. These all results were supported by linear shrinkage and bulk density of specimens.

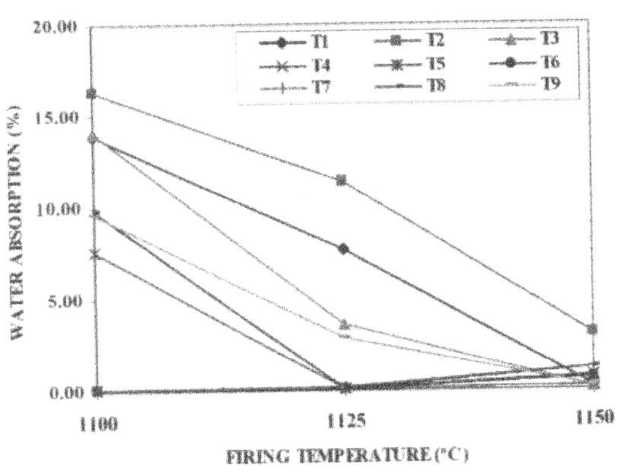

Fig3. Water absorption of specimens

According to DIN standard, water absorption of unglazed floor tile is not over than 5%. Therefore, fired specimens that were no shape distortion and water absorption lower than 5% were selected to examined phase and strength and chemical properties. Specimens T3, T4, T5, T6 and T7 were selected based on water absorption criteria for further examination.

Phase Analysis

XRD patterns show the presence of quartz (SiO_2), cristobalite (SiO_2), albite ($NaAlSi_3O_8$) and Anorthite ($Ca(Al_2Si_2O_8)$) peaks in all specimens. Rising zinc waste percentage, there was an decrease in the peak intensity of quartz and albite while the intensity of cristobalite and anorthite peaks were increased. The wollastonite ($CaSiO_3$) peak can be observed in specimens contain of 50% zinc waste or more (specimens T5, T6 andT7). An increase of zinc waste content in specimens leads to an increase of CaO content resulting in higher intensity of wollastonite peak.

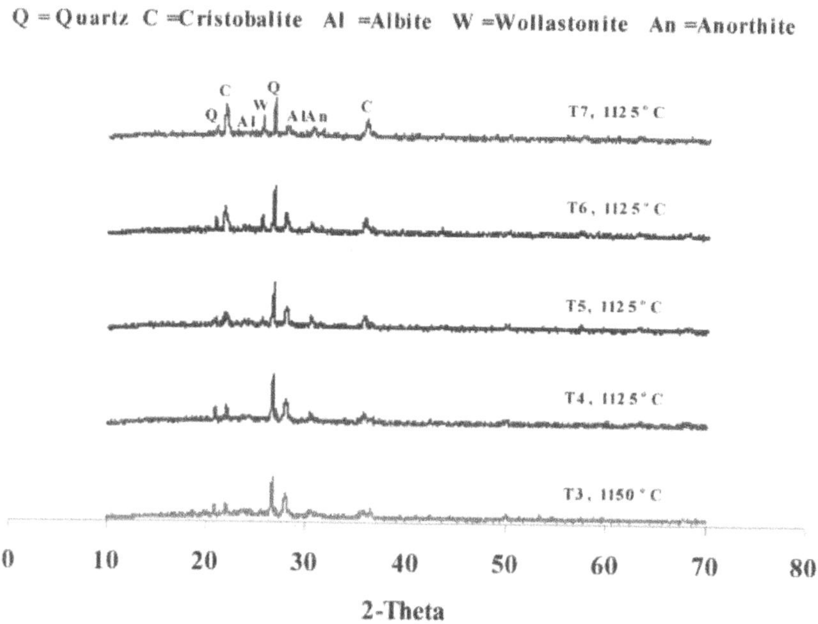

Fig4. XRD pattern of specimens

Mechanical Properties

Mechanical strength of fired specimens is shown in Fig5. Most of specimens gave the strength higher than the minimum limit of ISO 13006[8] (30 MPa for water absorption<3%), except T5 at 1150°C is 21.55 MPa. The maximum strength of specimens T5, T6 and T7 were obtained at lower temperature (1100°C) than T3 (1150°C) and T4 (1125°C). The strength of specimen T3 slightly increased as a function of firing temperature. The highest strength in this experiment was found in specimen T7 fired at 1100°C which is 107.26 MPa. The strength of specimens is strongly depending on density and water absorption of specimens

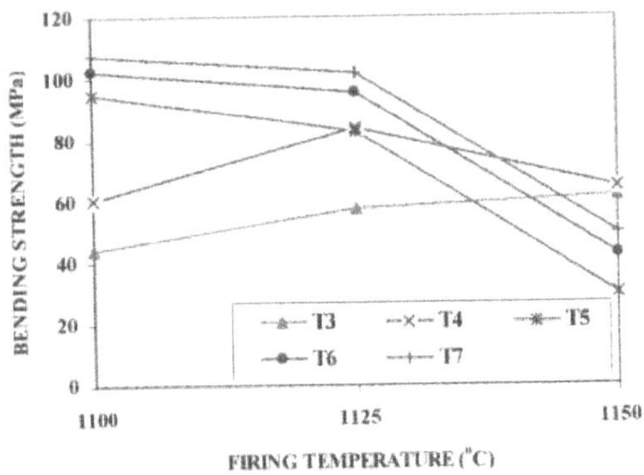

Fig5. Bending strength of specimens

Chemical Properties

Chemical resistance and toxic leaching of fired specimens with no shape distortion, good surface and water absorption obtained to zero are shown in TableIII.

TableIII. Chemical properties of specimens

Specimens	Firing temperature (°C)	Pb release(ppm)	Chemical resistance		
			HCl	KOH	HB pencil
T5	1100	1.51	fail	pass	pass
T6	1100	6.67	fail	fail	pass
T7	1100	4.21	pass	pass	pass
T4	1125	0.63	pass	pass	pass
T5	1125	0.87	pass	pass	pass
T6	1125	5.41	fail	pass	pass
T3	1150	0.06	pass	pass	pass

Considering to the chemical resistance, most of the specimens show good chemical resistance at any temperatures except T5, T6 fired at 1100°C and T6 fired at 1125°C. T5 fired at1100°C and T6 fired at 1125°C poor in acid resistance while T6 fired at 1100°C poor in both acid and base resistance.

The released Pb was examined by using TCLP method. The released Pb are varied from 0.06 ppm - 6.67 ppm. Pb release of 5 out of 7 specimens were lower than the limit of USA (5 ppm)[9], which are T5 fired at 1100°C and 1125°C, T7 at 1100°C, T4 at 1125°C and T3 at 1150°C.

The water absorption of all specimens was lower than 5%.The maximum strength of T3 and T4 is obtained when fired at 1150°C and 1125°C, respectively. Although the maximum strength of T5 and T6 are obtained at 1100°C, the chemical properties could be improved by firing at higher temperature 1125°C as can be seen in TableIII.

CONCLUSION

Reutilizing zinc waste as raw material for developing unglazed tile is one alternative to dispose of hazardous waste. Physical properties, mechanical property and chemical properties of specimens which contain zinc waste up to 70% are acceptable for this application. These properties are strongly dependent on firing temperature. At optimum firing condition of specimens T3 (1150°C), T4 (1125°C), T5 (1125°C), T6 (1100°C) and T7 (1100°C), bending strengths range from 61.6-107.26 MPa which is at least double of the strength of specimen with out zinc waste (30MPa). All of these specimens passed chemical and leaching test. These unglazed tiles can be prepared with high percentage zinc waste loading at lower temperature. This method is a lower cost method compares to vitrification process and very practical for normal tile production. The color of unglazed tiles can be obtained from zinc waste with no addition of expensive body stains.

ACKNOWLEDGEMENT

We are very grateful to Thailand graduate Institute of Science and Technology (TGIST), National Science and Technology Development Agency (NSTDA) Thailand for a financial support. We would like to thank Padeang Industry Public Co., Ltd. and Thai Ceramic Co., Ltd. for supporting raw materials.

REFERENCES

[1]M.Remero, J.Ma.Rincon, "Microstructureal characterization of a goethite waste from zinc hydrometallurgical process", *Materials letter*, 31 (1997), 67-73.

[2]L.Montanaro, N.Bianchini, J.Ma.Rincon, M.Romero,"Sintering behavior of press red mud wastes from zinc hydrometallurgy", *Ceramic international*, 27 (2001), 29-37.

[3] Vsevolod A.Mymrin, Haroldo A.Ponte, Patrico R.Impinnisi"Potential application of acid jarosite wastes as the main component of construction materials",*Construction and building materials*, 19 (2005), 141-146.

[4]S.Jiemsililers, B.Patarachao, P.Thavornniti, and S.Supothina, "Properties of glass-ceramics synthesized from hydrometallurgical, " *Melt Chemistry, Relaxation and Solidification Kinetics of Glasses*, Ceramic Transactions 170 (2005), 237-246.

[5]Standard test method for water absorption, bulk density, apparent porosity, and apparent specific gravity of fired whiteware products, ASTM Standard C373-88 (1998).

[6]Standard test method for resistance of ceramic tile to chemical substances, ASTM Standard C650-97 (1997).

[7]Method 1311 toxic characteristic leaching procedure(online).Available from: http://www.epa.gov/epaoswer/hazwaste/test/pdfs/1311.July, 1992

[8]Ceramic tiles-Definitions, classification, characteristics and marking, International standard ISO13006 (1998).

[9]M.Pelino, "Recycling of zinc-hydrometallugy wastes in glass and glass ceramic materials", *Waste Management*, 20 (2000), 561-568.

Eco-Materials Processing and Design
Edited by Koji Watari and Soo-Wohn Lee

DESIGN AND DEVELOPMENT OF ENVIRONMENTALLY-FRIENDLY PHOTOCATALYTIC ECO-MATERIALS USING SOLAR ENERGY

Masakazu ANPO* and Masaya MATSUOKA
Department of Applied Chemistry, Graduate School of Engineering
Osaka Prefecture University
E-mail: anpo@chem.osakafu-u.ac.jp

INTRODUCTION

Environmental pollution and destruction on a global scale, as well as the lack of sufficient clean and natural energy resources have drawn much attention and concern to the vital need for ecologically clean chemical technology, materials and processes--one of the most urgent challenges facing chemical scientists. Especially significant are photocatalytic processes that can make use of clean, safe, and abundant solar energy. In this regard, TiO_2 semiconductor materials are ideal and powerful photocatalysts due to their chemical stability, non-toxicity and high reactivity for the elimination of pollutants in air and water. However, titanium oxide photocatalysts can make use of only 3-4% of the solar irradiation, requiring the use of an UV light source such as a high pressure Hg lamp or black lamp.

As an approach in the development of visible-light responsive TiO_2 photocatalysts, various photosensitizing dyes adsorbed and/or supported on semiconducting catalysts were investigated. However, these photosensitizing dyes are not thermally and photochemically stable. Although numerous investigations have been carried out into visible light-responsive TiO_2 photocatalysts by adding small amounts of components such as cations and metal oxides, no significant results could be obtained and these initial trials were found to have limitations.

However, an advanced metal ion-implantation method was applied to modify the electronic properties of such TiO_2 photocatalysts by bombarding them with high energy metal ions. Thus, it was discovered that implantation with various transition metal ions such as V, Cr, Fe, Co and Ni by high voltage acceleration in the range of 50-200 keV could enable a large shift in the absorption band of these photocatalysts toward visible light regions, with different levels of effectiveness. Such a large shift allowed the metal ion-implanted TiO_2 photocatalysts to effectively and efficiently absorb solar light of up to 25-35 %. Furthermore, during the preparation of TiO_2 thin-film photocatalysts using various ion-engineering techniques such as an ion cluster beam method, a completely new technique, an RF-magnetron sputtering deposition method, was also discovered to modify the TiO_2 thin-film photocatalysts so that they could absorb and operate even under visible light irradiation. With this method, it was found that the photocatalytic reactivity of these TiO_2 thin-films and the wavelength regions of the absorption of visible light greatly depended on the sputtering conditions, such as the pressure of the Ar sputtering gas and the substrate temperature. This most recent investigation is significant in that effective visible light-responsive TiO_2 thin-film photocatalysts could be developed at low cost with the RF-magnetron sputtering deposition method, in contrast to the high cost of using the advanced ion-implantation method.

In this proceedings we report the results obtained by the successful application of advanced ion-beam techniques such as metal ion-implantation (Fig. 1-A), ionized cluster beam deposition

(Fig. 1-B), and RF magnetron sputtering deposition (Fig. 1-C) for the design and development of titanium oxide photocatalysts which are able to absorb and work not only under UV but also visible or solar light irradiation for the following different three systems: 1) TiO_2 nano-powdered materials; 2) Highly dispersed titanium oxide materials prepared within zeolite frameworks (Ti/zeolite); 3) TiO_2 thin film materials. These advanced physical methods were found to be one of the most suitable and promising approaches in the preparation of highly functional second-generation titanium oxide-based photocatalytic materials, leading to effective and efficient solar beam utilization of up to 20-30 %.

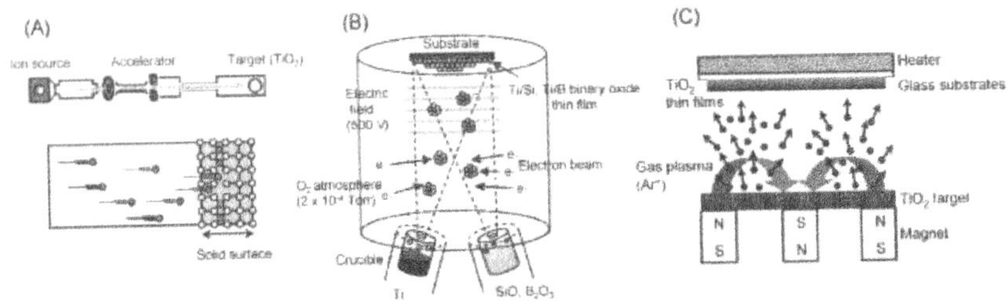

Fig. 1. The schematic diagrams of various ion-beam techniques.
(A) Metal ion-implantation method, (B) Ionized cluster beam deposition method, and
(C) RF-magnetron sputtering deposition method.

DESIGN AND DEVELOPMENT OF "SECOND-GENERATION TIO_2 PHOTOCATALYSTS" WHICH CAN OPERATE UNDER VISIBLE LIGHT IRRADIATION

Modification of the electronic state of TiO_2 by applying an advanced metal ion- implantation method

A highly advanced metal ion-implantation method was applied to modify the electronic properties of bulk TiO_2 photocatalysts by bombarding them with high-energy metal ions. It was discovered that such implantation with various transition metal ions, such as V, Cr, Mn, Fe, and Ni, accelerated by high voltage enabled a large shift in the absorption band of the titanium oxide catalysts towards the visible light region, with different levels of effectiveness. However, Ar, Mg, or Ti ion-implanted TiO_2 exhibited no shift in the absorption spectra, showing that such a shift is not caused by the high-energy implantation process itself but by some interactions between the transition metal ions and the TiO_2 catalyst. The absorption band of the Cr ion-implanted TiO_2 shifts smoothly towards the visible light region, the extent of the red shift depending on the amount and type of metal ions implanted, with the absorption maximum and minimum values always remaining constant. The order of effectiveness in the red shift was found to be as follows: V > Cr > Mn > Fe > Ni ions. Such a shift allows the metal ion-implanted titanium oxide to use solar irradiation more effectively, with efficiencies in the range of 20 - 30%.

Furthermore, such red shifts in the absorption band of the metal ion-implanted TiO_2 catalysts were observed for any kind of titanium oxide except the amorphous types, the extent of the shift different from sample to sample. It was also found that such shifts in the absorption band could be observed only after calcination of the metal ion-implanted TiO_2 samples in O_2 at around 723 -

823 K. Therefore, it is clear that calcination under O_2 atmosphere in combination with metal ion-implantation was instrumental in the red shift of the absorption spectrum toward visible light regions. These results clearly show that such shifts in the absorption band of the TiO_2 catalysts by metal ion-implantation are a general phenomenon and not a special feature of certain kinds of bulk TiO_2 catalysts.

The Cr ion-doped catalysts showed no shift in the absorption edge of TiO_2; however, a new absorption band appears at around 420 nm as a shoulder peak due to the formation of the impurity energy level within the band gap, its intensity increasing with the amounts of Cr ions chemically doped. These results indicate that the method of doping causes the electronic properties of the TiO_2 catalyst to be modified in completely different ways, thus confirming that only metal ion-implanted TiO_2 catalysts show such shifts in the absorption band toward visible light regions, even with much lower amounts of the ions, as compared with those for the chemically doped systems.

Neither with the unimplanted original TiO_2 nor with the chemically doped TiO_2 catalysts does the photocatalytic reaction proceed under visible light irradiation ($\lambda > 450$ nm). However, visible light irradiation of such advanced metal ion-implanted TiO_2 catalysts enabled them to initiate various significant photocatalytic reactions. Visible light irradiation ($\lambda > 450$ nm) of the Cr ion-implanted TiO_2 catalysts in the presence of NO at 275 K led to the decomposition of NO into N_2, O_2, and N_2O with a good linearity against the light irradiation time. Under the same conditions of visible light irradiation, the unimplanted original pure TiO_2 catalyst did not exhibit any photocatalytic reactivity. Moreover, the action spectrum for this reaction on the metal ion-implanted TiO_2 was in good agreement with the absorption spectrum of the catalyst, indicating that only metal ion-implanted TiO_2 catalysts were effective for the photocatalytic decomposition reaction of NO under visible light irradiation. Thus, it was found that metal ion-implanted TiO_2 catalysts could absorb visible light up to a wavelength of 400-600 nm and could thus operate effectively as photocatalysts: hence the name, "second-generation TiO_2 photocatalysts".

It is important to emphasize that the photocatalytic reactivity of the metal ion-implanted TiO_2 catalysts retained the same photocatalytic efficiency as the unimplanted original TiO_2 catalyst under UV light irradiation ($\lambda < 380$ nm). When metal ions were chemically doped into the TiO_2 catalyst, the photocatalytic efficiency decreased dramatically even under UV irradiation; this was due to the quick recombination of the photo-formed electrons and holes through the impurity energy levels formed by the doped metal ions within the band gap of the catalyst. These results clearly show that metal ions that are physically implanted do not work as electron and hole recombination centers, but work only to modify the electronic properties of the catalyst.

Various fieldwork experiments were conducted to test the photocatalytic reactivity of the newly developed TiO_2 catalysts under solar beam irradiation. Under outdoor solar light at ordinary temperatures, the Cr and V ion-implanted TiO_2 catalysts showed several times higher photocatalytic reactivity for the decomposition of NO than the unimplanted original TiO_2 catalysts. It was also found the V ion-implanted TiO_2 catalysts showed several times higher photocatalytic reactivity for the hydrogenation of C_3H_4 with H_2O than the original TiO_2. These results, together with the results obtained in the absorption spectra, clearly indicate that, by using second-generation titanium oxide photocatalysts developed by applying the metal ion-implantation method, the effective and efficient use of visible and solar light energy by photocatalysis is now possible.

The relationship between the depth profiles of the metal ions implanted within TiO_2 having the same amounts of metal ions, such as V or Cr ions, and their photocatalytic efficiency under visible light irradiation was investigated. It was found that, when the same amounts of metal ions were implanted into the deep bulk of the TiO_2 catalyst by applying high acceleration energy, the catalyst exhibited high photocatalytic efficiency under visible light irradiation. On the other hand, when low acceleration energy was applied, the catalyst exhibited low photocatalytic efficiency under the same conditions of visible light irradiation.

Increasing the number (or amounts) of metal ions implanted into the deep bulk of the TiO_2 catalyst was found to cause the photocatalytic efficiency to increase under visible light irradiation, passing through a maximum at around $6x10^{16}$ ions/cm^2, then decreasing with a further increase in the number of metal ions implanted. The presence of ions at the near surfaces could be observed by ESCA measurements only on samples implanted with a large amount of metal ions. These results clearly show that there are optimal conditions in the depth and amount of metal ions implanted to achieve high photocatalytic reactivity under visible light irradiation.

The ESR spectra of the V ion-implanted TiO_2 catalysts were measured before and after calcination of the samples in O_2 at around 723-823 K. Distinct and characteristic reticular V^{4+} ions were detected only after calcination at around 723-823 K. When a shift in the absorption band towards visible light regions by the V ion-implantation was observed, the presence of the reticular V^{4+} ions could be detected by ESR measurements. No such V ions having the same local structure or shifts in the absorption band could be observed with TiO_2 catalysts chemically doped with V ions.

The XANES and Fourier transforms of the EXAFS oscillations for the TiO_2 catalysts chemically doped and also physically implanted with Cr ions were measured in order to clarify the local structures of these Cr ions. Results obtained from the analyses of the XAFS spectra showed that, for the TiO_2 catalysts chemically doped with Cr ions by an impregnation or sol-gel method, the ions were present as aggregated Cr-oxides in octahedral coordination similar to Cr_2O_3 and in tetrahedral coordination similar to CrO_3, respectively. On the other hand, in the catalysts physically implanted with Cr, the ions were found to be present in a highly dispersed and isolated state in octahedral coordination, clearly suggesting that the Cr ions are incorporated in the lattice positions of the TiO_2 catalyst in place of the Ti ions.

These findings clearly show that the success in modifying the electronic state of TiO_2 by metal ion-implantation, enabling the absorption of visible light even longer than 550 nm, is closely associated with the strong and long-distance interaction which arises between the TiO_2 and the metal ions implanted, and not by the formation of impurity energy levels within the band gap of the catalysts. While the impurity energy levels are appeared due to the formation of aggregated metal oxide clusters, which are often observed in the chemical doping of metal ions and/or oxides.

These results obtained in the photocatalytic reactions and various spectroscopic measurements of the catalysts clearly indicate that the implanted metal ions are highly dispersed within the deep bulk of the TiO_2 catalysts and work to modify the electronic nature of the catalysts without any changes in the chemical properties of the surfaces. These modifications were found to be closely associated with an improvement in the reactivity and sensitivity of the TiO_2 photocatalyst, thus enabling the TiO_2 catalyst to absorb and operate effectively not only under UV but also under visible light irradiation. As a result, under outdoor solar light irradiation at ordinary temperatures, transition metal ion-implanted TiO_2 catalysts showed several times higher photocatalytic efficiency than the unimplanted original TiO_2 catalysts.

Visible light responsive Ti-oxide/zeolites

The XANES spectra of Ti/zeolite at the Ti K-edge clearly showed intense pre-edge peaks, which are related to the local structures surrounding the titanium oxide species in tetrahedral symmetry. The Fourier transforms of the EXAFS spectra for these Ti/zeolite catalysts exhibited a strong peak at around 1.83 Å, which could be assigned to the neighboring oxygen atoms, i.e., a Ti-O bond, indicating the presence of an isolated titanium oxide species. Moreover, curve-fitting analysis of the Fourier transforms for the EXAFS spectra showed that the titanium oxide species incorporated within the zeolite frameworks exist in tetrahedral coordination with a Ti-O bond distance of about 1.83 Å for various Ti/zeolite catalysts.

The absorption spectra of the V ion-implanted Ti/HMS and Ti/MCM-41 catalysts these catalysts at around 200–260 nm could be attributed to the charge transfer absorption process involving an electron transfer from the O^{2-} to Ti^{4+} ions of the highly dispersed tetrahedrally coordinated TiO_4 unit of these catalysts. These absorption spectra shift smoothly towards visible light regions, the extent strongly depending on the amount of V ions implanted. These results clearly suggest that the interaction of the implanted V ions with the TiO_4 units leads to the modification of the electronic properties of the titanium oxide species within the zeolite frameworks.

The V K-edge FT-EXAFS spectra of the Ti/HMS catalyst implanted with V ions indicate that the next-neighbours of the V environment are not the same as vanadium oxide based catalysts (e.g., V_2O_5) and suggests the formation of tetrahedral titanium oxides having a V-O-Ti bond instead of V-O-V linkages. These findings for the V-O-Ti bridge structures in the V ion-implanted Ti/HMS and also in the Ti/MCM-41 catalysts support the findings shown by the red shift in the absorption spectra for these catalysts.

The photocatalytic reactivity of the V ion-implanted Ti/ HMS and Ti/MCM-41 (and Ti/zeolite catalysts such as Ti-beta zeolite) catalysts for the decomposition of NO into N_2 and O_2 under visible light irradiation ($\lambda > 420$ nm) has been investigated. UV irradiation ($\lambda < 300$ nm) of Ti/HMS, Ti/MCM-41, and other Ti/zeolite catalysts in the presence of NO led to an efficient decomposition of NO into N_2 and O_2. However, as expected from their absorption spectra, the unimplanted original Ti/HMS, Ti/MCM-41, and other Ti/zeolites did not work to catalyze the photocatalytic decomposition of NO under visible light irradiation. Moreover, visible light irradiation of the V ion-implanted Ti/HMS, Ti/MCM-41, and other Ti/zeolite catalysts led to the efficient decomposition of NO into N_2 and O_2, and the yield of the photo-formed N_2 increased linearly with the irradiation time. These results clearly indicate that this decomposition reaction proceeds on the V ion-implanted Ti/HMS, Ti/MCM-41, and other Ti/zeolite catalysts under both UV and visible light irradiation. It was also confirmed that NO decomposition does not proceed at all under UV ($\lambda < 300$ nm) nor under visible light irradiation ($\lambda > 420$ nm) on the V ion-implanted HMS, MCM-41, and other zeolites. Also, the titanium oxide species in the Ti/zeolite, Ti/HMS, and Ti/MCM-41 catalysts were shown to be present in tetrahedral coordination within the zeolite frameworks. The implanted V ions and the highly dispersed Ti-oxide species form the Ti-O-V linkage, which leads to the modification of the electronic properties of the tetrahedral Ti-oxide species. These new properties enable the absorption of visible light to initiate reactions ($\lambda > 420$ nm) with high photocatalytic efficiency for the decomposition of NO into N_2 and O_2.

Application of a RF magnetron sputtering deposition method to design TiO_2 thin-film photocatalysts which operate under visible light irradiation

A radio-frequency magnetron (RF-MS) deposition method has recently been developed to prepare TiO_2 thin-film photocatalysts, using a TiO_2 plate as the sputtering target and Ar as the sputtering gas. Thus, an alternative and more practical method to prepare transparent TiO_2 thin-films which can initiate various significant photocatalytic reactions effectively even under visible light irradiation could be successfully performed. The UV-VIS absorption transmittance spectra of the TiO_2 thin- films prepared by the RF-MS deposition method at different substrate temperatures were measured. The thin-films prepared at relatively low temperatures ($T < 473$ K) exhibit high transparency and clear specific interference fringes in the visible light region, indicating that highly transparent and uniform thin-films have been formed on the substrate. The thin-films prepared at relatively high temperatures ($T > 773$ K) exhibit an efficient absorption in the visible light region, having a maximum absorption with the thin-film prepared at 873 K. In fact, these TiO_2 thin-films exhibited photocatalytic effectiveness for various reactions, such as the decomposition of NO as well as the oxidation of acetaldehyde with O_2, under both UV and visible light irradiation.

CONCLUSIONS

A summary of the preparation of nano-sized TiO_2 particles, highly dispersed titanium oxide species within zeolite cavities, titanium oxide based binary catalysts, and second-generation TiO_2 photocatalysts which can operate under visible light irradiation by an advanced metal ion-implantation method as well as by a new, more low-cost RF magnetron sputtering method has been presented. Detailed characterizations of these visible light-responsive TiO_2 thin-film photocatalysts were carried out at the molecular level, along with investigations into the various significant photocatalytic reactions that could be initiated.

The direct detection of the reaction intermediate species at the molecular level provided important information in explaining and clarifying the reaction mechanisms behind the observed photocatalytic reactions as well as the electronic properties of the photocatalyst surface. Furthermore, the advanced metal ion-implantation method and the RF-MS deposition method have opened the way to many innovative possibilities in the utilization of solar or visible light energy by realizing the development of unique second-generation TiO_2 photocatalysts and visible light-responsive molecular-sized titanium oxide species incorporated within the cavities and frameworks of zeolites and mesoporous molecular sieves. A combination of these fascinating and unique zeolite-incorporated systems and advanced ion-engineering techniques will provide new approaches in the utilization of solar energy as the most abundant and safe energy source, significantly to address environmental pollution on a large global scale as well as realizing the production of H_2 in the photocatalytic splitting of H_2O by solar irradiation.

ACKNOWLEDGMENTS

The author thanks his colleagues, Drs. H. Yamashita, M. Takeuchi and N. Zhanpeisov and PhD students (Drs. Y. Ichihashi, M. Harada, S. Higashimoto, K. Ikeue, S.W. Joo, Y. Shioya, and S. Dohshi) as well as many graduate-school students and post-graduate students for their cooperation.

REFERENCES

1) M. Anpo, M. Takeuchi, H. Yamashita, S. Kishiguchi, A. Davidson and M. Che *"Semiconductor Photochemistry and Photophysics"*, Marcel Dekker, Inc. (2003), p. 283.
2) Dohshi, M. Takeuchi, and M. Anpo, *J. Nanosci. Nonotech.*, 1, 337 (2001).
3) M. Anpo, S. Dohshi, and M. Takeuchi, *J. Cermic Process. Res.*, 3, 258 (2002).
4) S. Dohshi, M. Takeuchi, and M. Anpo, *Catal. Today*, 85, 199 (2003).
5) H. Yamashita, M. Harada, A. Tanii, M. Honda, M. Takeuchi, Y. Ichihashi, and M. Anpo, *Stud. Surf. Sci. Catal.*, 130, 1931 (2000).
6) M. Takeuchi, M. Matsuoka, H. Yamashita, and M. Anpo, *J. Synchrotron Rad.*, 8, 643 (2001).
7) M. Matsuoka and M. Anpo, *J. Photochem. Photobiol. C: Photochem. Rev.*, 3, 225 (2003).
8) M. Anpo, M. Takeuchi, K. Ikeue, and S. Dohshi, *Current Opinion Sold State Mater. Sci.*, 6, 381 (2002).
9) Y. Ichihashi, H. Yamashita, and M. Anpo, *Kinou-Zairyou (Functional Materials)*, 16, 12 (1996).
10) M. Anpo, H. Yamashita, and Y. Ichihashi, *Optronics*, 186, 161 (1997).
11) M. Anpo, *Catal. Survey Jpn.*, 1, 169 (1997).
12) M. Anpo, Y. Ichihashi, M. Takeuchi, and H. Yamashita, *Res. Chem. Intermed.*, 24, 143 (1998).
13) M. Anpo, Y. Ichihashi, M. Takeuchi, and H. Yamashita, *"Sci. Tech. in Catal. 1998"*, eds. by H. Hattori and K. Otsuka, Kodan-sha, Tokyo (1999), p. 305.
14) M. Anpo, M. Takeuchi, H. Yamashita, T. Hirao, N. Itoh, and N. Iwamoto, *Proc. Intern. Conf. "The 4th International Conference ECOMATERIAL"*, (Gifu) (1999), p. 333..
15) M. Anpo, Y. Ichihashi, M. Takeuchi, and H. Yamashita, *Stud. Suf. Sci. Catal.*, 121, Proc. Tocat-3, Tokyo, 305 (1999).
16) M. Anpo, M. Takeuchi, S. Kishiguchi and H. Yamshita, *Surf. Sci. Jpn.*, 20, 60 (1999).
17) M. Anpo, H. Yamashita, S. Kanai, K. Sato, and T. Fujimoto, US patent No. 6,077,492 (June 20, 2000).
18) M. Takeuchi, H. Yamashita, M. Matsuoka, T. Hirao, N. Itoh, N. Iwamoto, and M. Anpo, *Catal. Lett.*, 66, 185 (2000).
19) M. Takeuchi, H. Yamashita, M. Matsuoka, T. Hirao, N. Itoh, N. Iwamoto, and M. Anpo, *Catal. Lett.*, 67, 135 (2000).
20) M. Anpo, *"Green Chemistry"*, Oxford University Press (2000), p. 1.
21) M. Anpo, "*Pure Appl. Chem.* IUPAC", 72, 1265 (2000).
22) M. Anpo, *Stud. Surf. Sci. Catal.*, 130, 157 (2000).
23) M. Anpo, S. Kishiguchi, Y. Ichihashi, M. Takeuchi, H. Yamashita, K. Ikeue, B. Morin, A. Davidson, and M. Che, *Res. Chem. Intermed.*, 27, 45, 459 (2001).
24) M. Anpo, Keynote Lecture, Studies in Surface Science and Catalysis, 130, *"12th International Congress on Catalysis"*, ed. by A. Corma, F. V. Melo, S. Mendioroz, and J. L. G. Fierro, Elsevier (2000), p.157.
25) M. Anpo and M. Takeuchi, *Inter. J. Photoenergy*, 3, 1 (2001).
26) M. Takeuchi, M. Anpo, T. Hirao, N. Itoh, and N. Iwamoto, *Surf. Sci. Jpn.*, 22, 9, 561 (2001).
27) H. Yamashita, M. Harada, J. Misaka, M. Takeuchi, Y. Ichihashi, F. Goto, M. Ishida, T. Sasaki, and M. Anpo, *J. Synchrotron Rad.*, 8, 569 (2001).
28) H. Yamashita, M. Harada, J. Misaka, M. Takeuchi, K. Ikeue, and M. Anpo, *J. Photochem. Photobiol., A: Chem.*, 148, 257 (2002).
29) M. Anpo and M. Takeuchi, *J. Catal.*, 216, 505 (2003).
30) M. Anpo and M. Takeuchi, *"Handbook of Ion Engineering"*, Institute of Ion-Engineering, Osaka (2002), p. 943.
31) M. Takeuchi, M. Anpo, T. Hirao, N. Itoh, and N. Iwamoto, *Surf. Sci. Jpn.*, 22, 9, 561 (2001).
32) M. Anpo, *Bull. Chem. Soc. Jpn.*, 77, 1427 (2004), and references therein.

Eco-Materials Processing and Design
Edited by Koji Watari and Soo-Wohn Lee

DEGRADATION OF BENZENE USING PHOTOCALYTIC TiO_2-BASED NANO-COMPOUNDED COATINGS

Huang Chen, Soo Wohn Lee
Department of Materials Engineering, Sun Moon University
Asan, ChungNam, South Korea, 336-708

ABSTRACT

TiO_2 photocatalysis is considered to be effective for air purifying, even their concentrations of the pollutants that pose serious threat to health or comfort are quite low. This work was designed to study on degradation of benzene in gas-state using TiO_2 coatings as photocatalysis. In order to enhance the photocatalytic activity of TiO_2 coatings, 10wt. % nano-sized ZnO, CeO_2 and SnO_2 were added into nano-TiO_2 powder (P25), respectively. The mixture powders were milled and evaporated with a rotary evaporator. In this work, three kinds of TiO_2-based nano-compounded coatings were successfully prepared by plasma spraying technology using the evaporated powders as feedstock. Using the photocatalytic degradation of gas-state benzene as a reaction model, their photocatalytic performances of the three plasma sprayed TiO_2-based nano-compounded coatings were evaluated comparably. It was found that the plasma sprayed TiO_2-based nano-compounded coatings could almost degrade 100ppm benzene in 10h. The effects of their phase composition, surface morphology as well as surface chemical state on photocatalytic performances of the plasma sprayed TiO_2-based nano-compounded coatings will be discussed also.

INTRODUCTION

Accompanying the fast development of global economy in the past several decades, environmental pollution has becoming more and more serious. Recently, photocatalytic technology is held the highest hopes to resolve the problem. Titanium dioxide (TiO_2) is hotspots of this research, and a large number of applications of TiO_2 in photocatalytic technology has been developed, such as self-cleaning, air cleaning, water purification and self-sterilizing [1].

For photocatalytic materials, specific surface area is crucial characterization. The experimental results have demonstrated the photocatalytic performance is largely improved by using finer TiO_2 particles, increasing TiO_2 content and making porous structures [2]. In photocatalytic water treatment, the difficulty in recovering TiO_2 from treated water is a major obstacle. To solve this problem, two commonly immobilization methods are bonding process with binder and film process. In this work, three kinds of evaporated TiO_2-based nano-compounded powders were immobilized on foamed aluminum plates by plasma spraying technology.

EXPERIMENTAL PROCEDUCES

Compounded semiconductor is an effective method to enhance photocatalytic activity by increasing separation of photogenerated electron-hole pairs as well as broadening wavelength of irritation light. In this work, 10wt.% ZnO, CeO_2, SnO_2 powders were added into the TiO_2 powder (P25, 20-50nm, Degussa, Germany) respectively. Using alcohol as the medium, the three TiO_2-based nano-compounded powders were mixed and milled for 24h, and then evaporated by the rotary evaporator (R-114, Büchi Labo Ratoriums Technik AG., Switzerland). All the three agglomerated powders are irregular and loose. They are in the range of 1-30μm. To increase contact area with pollutants, foamed aluminum plate with dimensions of 30×20×1cm^3 was used as the substrate which possesses pores with 2-8mm in diameter. TiO_2-based coatings were deposited on the largest face of the foamed substrate. A-2000 atmospheric plasma spraying equipment (F4-MB gun, Sulzer Metco AG, Switzerland) was applied to deposit the TiO_2-based nano-compounded photocatalytic coatings with evaporated powders as feedstock. Table 1 lists the spraying parameters used in this work.

Table 1 Spraying parameters for the photocatalytic coatings

Argon flow rate (slpm)	40
Hydrogen flow rate (slpm)	9
Carrier gas flow rate (slpm)	3.5
Spraying power (kW)	36.9
Spraying distance (mm)	150

An X-ray diffractometer (Rigaku, Tokyo, Japan) operating with CuKα (λ=1.54056Å) radiation was applied to identify phase compositions. Morphologies of coatings were observed with a Field Emission Scanning Electron Microscope (JSM-6700F, Shimadzu, Tokyo, Japan). Surface chemical states of coatings were studied using X-ray photoelectron spectroscopy (Escalab 220-IXL). Their photocatalytic activities of the three plasma-sprayed TiO_2-based nano-compounded coatings were evaluated comparably by degradation of 100ppm gas-state benzene with a discharge photoelectron catalytic system [3].

RESULTS AND DISCUSSION

XRD spectra demonstrate that the dominant phase is still anatase phase for the three evaporated TiO_2-based powders as shown in Fig. 1. However, during the plasma spraying, anatase phase in the feedstock transformed to rutile phase as evidenced by XRD spectra of the three plasma-sprayed TiO_2-based nano-compounded coatings (Fig. 2). Compared with the

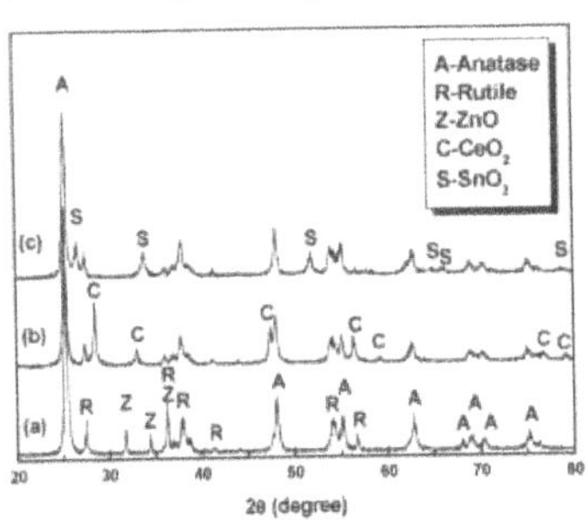

Fig. 1. XRD spectra of the evaporated powders: (a) 90%TiO_2-10%ZnO, (b)90%TiO_2-10%CeO_2 and (c) 90%TiO_2-10%SnO_2.

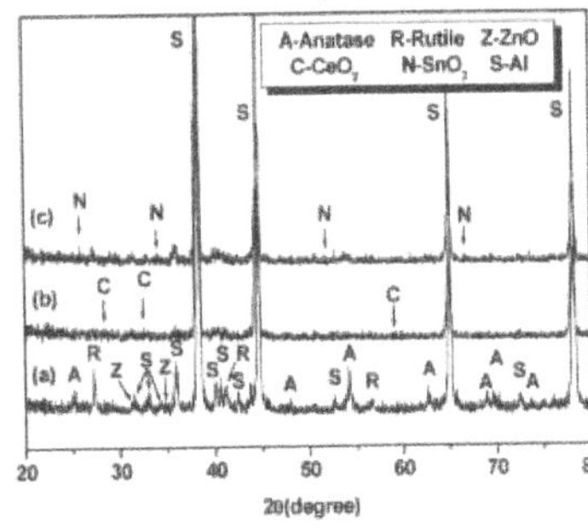

Fig. 2. XRD spectra of the nano-compounded coatings: (a) 90%TiO_2-10%ZnO, (b)90%TiO_2-10%CeO_2 and (c) 90%TiO_2-10%SnO_2.

feedstock, all of the anatase contents in the three as-sprayed coatings decreased after plasma spraying. It's ascribed to the extreme high temperature of plasma spraying process. Especially for the 90%TiO_2-10%ZnO powder, its anatase content reduced remarkably from 81.6% to 31% in the as-sprayed coating as plotted in Fig. 3. As can be seen from Fig. 3, the plasma-sprayed 90%TiO_2-10%CeO_2 coating has the highest anatase content, it is high up to 45.9%. It is 39% for in the case of 90%TiO_2-10%SnO_2 coating. As discussed later, anatase content is an important factor for photocatalytic activity. Moreover, it is noticeable that several peaks with higher relative intensity of the substrate (Al plate) appear in the XRD spectra. This may be explained on the base of thin thickness of the plasma-sprayed TiO_2-based nano-compounded coatings.

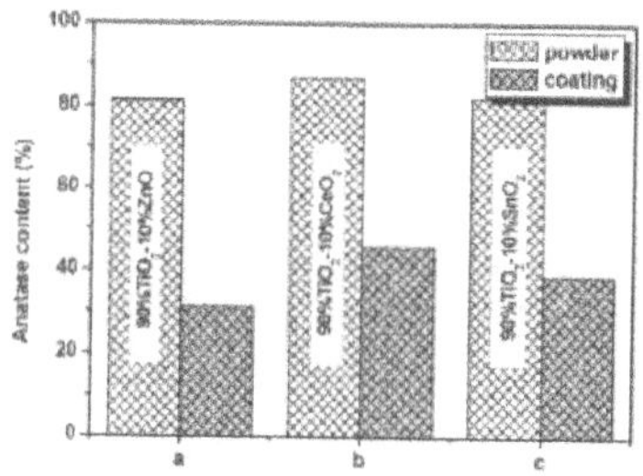

Fig. 3. Phase compositions of TiO_2-based powders and coatings: (a) 90%TiO_2-10%ZnO, (b) 90%TiO_2-10%CeO_2 and (c) 90%TiO_2-10%SnO_2.

Figure 4 shows their surface morphologies of the plasma-sprayed TiO_2-based nano-compounded coatings on foamed aluminum plates deposited from the three evaporated powders. All the TiO_2-based coatings were bonded well to the substrates. However, the coating thickness is about several micrometers and quite heterogeneous. The typical characterization of plasma-sprayed coatings using evaporated powder as feedstock is rough surfaces with melted splats and non-melted or partially melted particles. The melted splats have good flattening, while the non-melted or partially melted particles display ball-like shape resembling the starting nano-particle. On the splats' surfaces of the as-sprayed coatings, fine nano-crystalline particles were observed which could provide reactive sites for photocatalytic reaction.

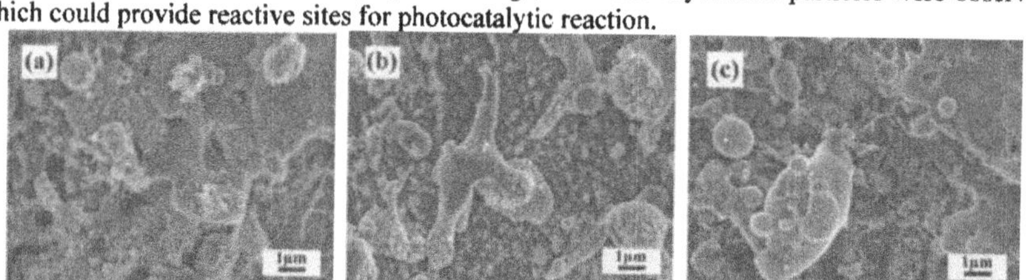

Fig. 4. FESEM morphologies of the plasma-sprayed TiO_2-based nano-compounded coatings:(a) 90%TiO_2-ZnO coating, (b)90%TiO_2-10%CeO_2 coating and (c)90%TiO_2-10%SnO_2 coating.

XPS analysis was used to testify chemical state of the as-sprayed TiO_2-based nano-compounded coatings in this work. Fig. 5 presents the evaluation of O1s for the three plasma-sprayed TiO_2-based nano-compounded coatings. Clearly, the left side of all the O1s XPS is wider than the right. By resolving their XPS curves, it was found that they were composted of two or three peaks as showed in Fig. 5. Peak 1 is believed to be connected with oxygen in the crystal lattice. Peak 2 is assigned to surface hydroxyl and peak 3 is due to physical-adsorbed oxygen. In the case of plasma-sprayed 90%TiO_2-10%CeO_2 coating, peak 3 is absent. The reason is still not clear. The surface hydroxyl group plays an important role in photocatalytic reaction [2].

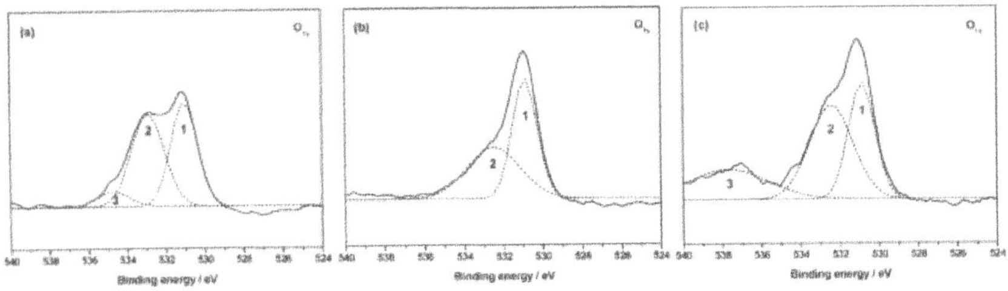

Fig. 5. XPS spectra of O1s recorded on the surfaces of plasma sprayed TiO_2-based nano-compounded coatings: (a) 90%TiO_2–10%ZnO, (b) 90%TiO_2–10%CeO_2 and (c) 90%TiO_2–10%SnO_2.

Using the photocatalytic degradation of gas-state benzene as the reaction model, their photocatalytic activities of the three plasma sprayed TiO_2-based nano-compounded coatings were evaluated comparably. For comparison, a blank test, the foamed aluminum plate without any coating was performed simultaneously. The dependence of benzene concentration on irradiation time using the three plasma-sprayed TiO_2-based nano-compounded coatings as photocatalyst was showed in Fig. 6. It's noted that the three plasma-sprayed TiO_2-based nano-compounded coatings exhibited similar photocatalytic activity, they almost degraded 100ppm gas-state benzene in 10h as can be seen from Fig. 6. Approximate 97.1% of benzene was degraded when used the plasma-sprayed 90%TiO_2-10%ZnO coating as the photocatalyst. It is slightly lower than the other two kinds of plasma-sprayed TiO_2-based nano-compounded coatings.

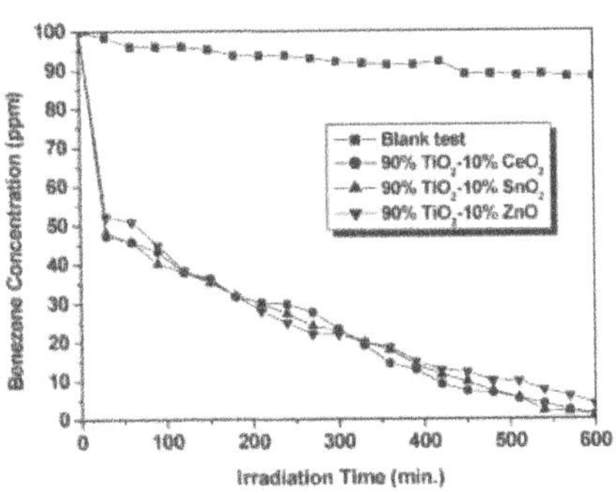

Fig. 6. The dependence of benzene concentration on irradiation time: (a) 90%TiO_2-ZnO coating; (b) 90%TiO_2-10%CeO_2 coating; (c) 90%TiO_2-10%SnO_2 coating.

Photocatalytic degradation of gas-state benzene is a complex process. Many factors could affect the final degradation rate. It's reported that high anatase content is better to show high photocatalytic activity [2,4]. Moreover, it has been accepted widely that the mixture of anatase and rutile phases has higher photocatalytic activity compared to pure anatase alone [2]. In this work, the plasma-sprayed 90%TiO_2-10%CeO_2 coating has the highest anatase content and exhibits better photocatalytic activity in degradation of gas-state benzene. In additional, surface morphology of coatings is important for photocatalytic activity, because photocatalytic reaction usually occurs on the surface of photocatalyst. As mentioned before, the surfaces of plasma-sprayed TiO_2-based nano-compounded coatings with evaporated powders as the feedstock exhibit rough structure and lots of nanoparticles appear on their surfaces, which could provide high surface area and active sites for the photocatalytic reaction of degradation benzene.

Surface hydroxyl (peak 2) also plays an important role in photocatalytic reaction [5]. The photogenerated holes can attack the surface hydroxyl and yield surface-bound hydroxyl radical ($^{\cdot}OH$), which is an oxidizing agent. Hence, with the increasing of surface hydroxyl, the chance to form hydroxyl radical with photogenerated holes is increased. Correspondently, the transfer of photogenerated holes toward surface is accelerated resulting in the improvement of photocatalytic activity. That's the reason why the plasma-sprayed 90%TiO_2-10%SnO_2 coatings exhibited almost the same photocatalytic activity, although its anatase content is lower than the 90%TiO_2-10%CeO_2 coating.

CONCLUSION

It is feasible to prepare photocatalytic coating by plasma spraying technology using evaporated powder as feedstock. The plasma-sprayed TiO_2-based nano-compounded coatings on foamed aluminum plates almost degraded 100ppm gas-state benzene in 10h. Their photocatalytic activity is dependent on anatase content, surface morphology and hydroxyl content.

REFERENCES

[1]A. Fujishima, K. Hashimoto, T. Watanabe, "TiO_2 Photocatalysis Fundamentals and Applications," BKC, Inc., Tokyo, 1999.

[2]J. A. Byrne, B. R. Eggins, N. M. D. Brown, B. Mckinney and M. Rouse, "Immobilization of TiO_2 Powder for the Treatment of Polluted Water," *Appl. Catal. B: Environ.*, 17, 25–36 (1998).

[3]H. S. Kim, E. A. Lee, C. H. Han, et al., "Discharged Photoelectrocatalytic System for the Enhancement of Photocatalytic Performance," *Materials Science Forum*, **436**, 205-209 (2003).

[4]G. Balasubramanian, D. D. Dionysiou, M. T. Suidan, I. Baudin, and J. M., Laine, "Evaluating the Activities of Immobilized TiO_2 Powder Films for the Photocatalytic Degradation of Organic Contaminants in Water," *Appl. Catal. B: Environ.*, **47**, 73–84 (2004).

[5]E. Pelizzetti, and C. Minero, "Mechanism of the Photo-oxidative Degradation of Organic Pollutants over TiO_2 Particles," *Electrochim. Acta*, **38(1)**, 47–55 (1993).

Eco-Materials Processing and Design
Edited by Koji Watari and Soo-Wohn Lee

UNDERSTANDING THE ADSORPTION AND PHOTOREACTION OF OXALIC ACID ON ANATASE NANOPARTICLES

Cecilia B. Mendive[1,2*], Thomas Bredow[3], Miguel A. Blesa[2] and Detlef W. Bahnemann[1]

[1]Institut für Technische Chemie, Universität Hannover, D-30167 Hannover, Germany.
[2]Unidad de Actividad Química, Centro Atómico Constituyentes, Comisión Nacional de Energía Atómica, 1650 San Martín, Provincia de Buenos Aires, Argentina.
[3]Theoretische Chemie, Universität Hannover, D-30167 Hannover, Germany.

ABSTRACT

The adsorption and photoreaction of oxalic acid on the surface of TiO_2 anatase nanoparticles is studied employing a combined experimental and theoretical approach. In the dark the adsorption reaches an equilibrium state as a mixture of adsorbed water and oxalic acid molecules with the latter forming two different surface complexes. When the system is illuminated with UV(A) light the surface is enriched in adsorbed oxalic acid by a replacement of the adsorbed water molecules, favouring one of the adsorbed structures.

INTRODUCTION

The system under consideration consists of TiO_2 nanoparticles in contact with an aqueous solution of oxalic acid. One driving force for pursuing this research project concerning the TiO_2 surface is the wide range of its applications[1 - 11] and the expectation that more insight into surface properties on the fundamental level will help to improve materials and device performance in many fields.

Titanium dioxide is a preferred system for experimentalists because it is well suited for many experimental techniques. However, TiO_2 is also a preferred system for many applications: it is used in heterogeneous catalysis, as a photo-catalyst, in solar cells for the production of hydrogen and electrical energy, as a gas sensor, as a white pigment (e.g. in paint and cosmetics products), as a corrosion-protective coating, as an optical coating, in ceramics, and in electrical devices. Titanium dioxide is also important in earth science, plays a role in the biocompatibility of bone implants, is being discussed as a gate insulator for the new generation of MOSFETS and as a spacer material in magnetic spin-valve systems, and finds applications in the nanostructured form in Li-based batteries and electrochromic devices.

By far, the most actively pursued applied research on titania concerns its use for the photo-assisted degradation of organic molecules. The applications of this process range from purification of wastewaters, disinfection based on the bactericidal properties of TiO_2[5] (for example, in operating rooms of hospitals), use for self-cleaning coatings on car windshields[6], to protective coatings of marble[7] (for the preservation of ancient Greek statues against environmental damage). It was even shown that subcutaneous injection of a TiO_2 slurry in rats,

* Corresponding author e-mail: hansmann@iftc.uni-hannover.de

and subsequent near-UV illumination, could slow or even revert the development of tumor cells[8 - 10].

We believe that a better understanding of the underlying mechanistic details will serve the improvement of these catalytic reactions and this is one of the main motivations of this investigation of the surface of TiO_2 particles.

On the other hand, the choice of oxalic acid is principally based on three important points: 1) its simplicity as a molecule which leads to its classification as a primary model compound, 2) the fact that it is in many cases one of the last products of degradation of more complex compounds and 3) its high relevance for biological systems, since, for example, it is a member of the citric acid cycle of the cell respiration and it is present in nature in every living organism and also in their environment as a metabolic or anabolic product.

This work presents the results of a systematic investigation of the TiO_2-oxalic acid system which involves both, experimental work and quantum-chemical calculations.
Concerning the experimental work, a series of experiments has been carried out comprising of at least three different steps:

Measurements of the adsorption kinetics in the dark (i.e., in the absence of UV-illumination). In this initial stage, one or more surface complexes are formed and their evolution can be monitored as a function of time by means of the FTIR-ATR (Fourier Transformed Infrared Spectroscopy - Attenuated Total Reflection) in situ technique.

Measurements of the dynamic equilibrium between the TiO_2 surface and the oxalic acid from the aqueous solution. Surface and bulk concentration of the organic acid are measured and compared in order to obtain informations concerning the number of active sites available for oxalate on different commercial TiO_2 products which differ in their BET surface area or their crystalline modification. Both the FTIR-ATR in situ technique and analytical measurements employing high performance liquid chromatography (HPLC) have been employed.

Measurements of the changes that occur to the adsorbed oxalic acid under steady-state UV illumination of the TiO_2 surface. Several photochemical reactions (photo-degradation, desorption and structure interconversion) take place and possible mechanisms can be suggested in order to explain the results. At this stage the implementation of the FTIR-ATR in situ technique gained extreme importance.

As a complementary tool and to challenge the hypotheses generated from the analysis of the experimental data, quantum chemical calculations have been carried out. We have focused on three different topics:

Determination of the most representative adsorption structures of the system. Geometry optimization of the adsorbants (oxalic acid and water, since they are both present in the system and will compete for adsorption sites on the TiO_2 surface) and of the TiO_2 surface

Simulation of the FTIR spectra within the desired window of wavenumbers employed in the experimental work.

It is intended to clarify some conflicting results concerning the adsorption of the test molecule[12, 14, 15, 18], and the changes that occur upon UV illumination of the system TiO_2/oxalic acid. Moreover, it can be anticipated that the insights obtained with this system can be transferred, at least in part, to other systems[13, 16, 17, 19]; and that the understanding thus gained will also be useful for the improvement and future development of application methods that in relationship with our environment appear to become more necessary in every moment.

EXPERIMENTAL PROCEDURE

Materials

The titanium dioxide powder used throughout this study has been obtained from Kemira. This commercial product, S-230, has been selected as being representative for the anatase phase, with 230 $m^2.g^{-1}$ BET surface area. All other reagents were of analytical grade and used as received. All solutions were prepared with MilliQ water from a SARTORIUS ARIUM 611 apparatus (conductance = 18.2 MΩ cm^{-1}).

The equipment

As mentioned in the introduction, the FTIR-ATR in situ technique has been employed [12, 14] together with additional standard techniques of HPLC (High Performance Liquid Chromatography) in order to perform the measurements.

The FTIR-*ATR in situ* spectra were recorded at approx. 20°C, using a BOMEM MB 122 instrument equipped with a liquid N_2 cooled MCT-A (mercury-tellurium-cadmium) detector. A horizontal Pike Technologies ZnSe – ATR unit 45° was used. The interferometer and the infrared light path in the spectrometer were constantly purged with Argon to avoid H_2O and CO_2 contamination in the spectra which only lead to a lowering of the total energy output for the detector and thus to a slight decrease of the method ′s sensitivity.

The final spectral resolution was 4 cm^{-1} since every scan is a double scan of two simple scans with a resolution of 8 cm^{-1} (from low frequencies to higher ones and in the opposite direction) interlaced by the equipment. Each final spectrum is the average of 250 scans. Base line corrections were made in order to eliminate minor fluctuations due to instrumental instabilities. The spectra were not smoothed and no ATR corrections were made.

The lamps used were two UV-A tubes (PHILIPS CLEO 15W), which have a maximum of emission at wavelengths between 300 and 400 nm. The energy output of the lamps employed in this work was 0.2 $mW.cm^{-2}$ and it was checked daily employing an UVA-meter from Dr. Hönle/UV-Technology. The semi-micro pH-electrode used was a Thermo-Orion Ross 8115 combined with a reference electrode Ag/AgCl. The automatic dosing unit employed was from Metrohm, a Dosimat 655 coupled with an Impulsomat unit. The High Performance Liquid Chromatograms were obtained employing an ECOM Spol.SR.O. instrument with an UV detector and autosampler from Basic Marathon. The column employed for the quantification of the oxalic acid aqueous solutions, was a NC04 (250 x 4.0 mm) Waters-Spherisorb ODS2 5.0 μm Bischoff Cromatography. The wavelength selected for the detector of the chromatograph was 210 nm. The mobile phase was H_3PO_4 0.01% v/v. The temperature in the compartment of the separation column has been kept constant at 30°C.

METHODS

The set-up

The ATR unit has been modified by means of the addition of an attached closed compartment, transparent to the incident UVA light, to the surface of the ZnSe crystal. This ensures that the solution can flow continuously through the system and several parameters such as pH, ionic strength and concentration of the bulk adsorbant can be controlled and regulated.

In the ATR unit, the ZnSe crystal has a truly central role[12, 19]. It is placed in the pathway of the IR light to ensure that the latter must pass through the crystal before arriving at the detector. Inside the crystal the light is reflected several times between the internal upper and lower faces. The number of reflections depends on the exact geometrical dimensions of the crystal. Relevant for the method itself is that, in every reflection, part of the light escapes from the crystal as an evanescent wave probing what has been placed on its top.

Thus, two main advantages of the ATR technique are obvious. On the one hand, the multiple reflections lead to stronger signals (a better ratio signal/noise). On the other hand, it is possible to avoid the undesired effects caused by the dispersion of the light as it crosses a liquid medium, which in the context of this work would be, the bulk aqueous solution or the TiO_2 suspension.

For this circulating unit 30 ml of circulating solution are employed. The oxalic acid concentration was chosen as 2×10^{-3} $mol.dm^{-3}$ and was maintained in circulation with a flow of approx. 4 $ml.min^{-1}$. This concentration of oxalic acid was chosen to ensure that it cannot be detected by the FTIR/ATR technique employed here. Thus, every detected signal can be attributed only to the adsorbed molecules.

The pH of the circulating solution was fixed and kept carefully constant at 3.70 ± 0.02 with a programmed dosing unit. The ionic strength has been adjusted to 0.01 $mol.dm^{-3}$ with KCl, and during the whole experiment the solution was constantly purged with O_2.

The TiO_2 particles layer

A thin layer of immobilized TiO_2 particles was deposited for each experiment on the surface of the ATR ZnSe crystal in the following way: a TiO_2 suspension was prepared at a concentration of 5.75 $g.L^{-1}$ and dispersed by sonication for 15 min in an ultrasonic cleaning bath. An aliquot of 200 μL of this suspension was placed on one corner of the surface of the crystal together with 200 μL of distilled water. This small volume was simply spread by balancing the unit gently and manually a few times, in order to evenly cover the surface of the crystal. This thin layer of suspension was directly evaporated to dryness at room temperature overnight. This final layer of particles thus obtained appears very homogeneous under visual inspection and has a thickness of 1.7 ± 0.3 μm (AFM measurements[18]), equivalent to a coverage of 2.3 $g.m^{-2}$. Thus, employing these conditions the evanescent wave of infrared light can probe the film properly to obtain the desired information concerning the species formed at the surface of the particles.

The layers thus prepared remain stable over the entire course of the experiment. Thus, it can be assumed that the effective path-lengths at all wavelengths remain unchanged[18]. It is also assumed here that the layer is isotropic, since the particles are small and randomly orientated, and uniformities in the layer thickness are averaged over the area of the ATR plate.

Adsorption and photo-catalytic experiments

The TiO_2 – coated crystal was equilibrated with a circulating oxalic acid–free aqueous solution for 30 minutes; during this period, sequential spectra were collected at intervals of 128 sec (approx. 2 min). When the last spectrum had been recorded, the UV lamp was turned on and another sequence of spectra was recorded during 30 minutes at the same time intervals. These two groups of spectra were considered as the blank reference spectra in the dark and under illumination, respectively. The illuminated blank spectra were recorded in order to have a pre-illumination step before the TiO_2 particles were brought in contact with the oxalic acid solution. Despite of collecting the blank reference spectrum under illumination, this procedure was just carried out with the purpose of checking and controlling the system.

Immediately after the blank reference spectra were recorded, the circulating solution was replaced by the test solution containing the desired amount of oxalic acid and a new set of spectra was collected in a similar way, but with the difference that 2 hours of equilibration in the dark and 8 hours for the measurements under illumination were employed.
The final spectra were obtained by subtracting the corresponding reference, which corresponds to the last spectrum of the blank in the dark and under illumination.
Aliquots of the circulating reference blank and the oxalic acid solution, respectively, were taken for HPLC analysis employing the following steps:

a) At the beginning, before any contact with the TiO_2 layer
b) After equilibration with the TiO_2 layer
c) After the illumination period

THEORETICAL PROCEDURE

The quantum chemical calculations - MSINDO

Quantum chemical calculations have been performed using MSINDO (Modified Symmetrically Orthogonalized Intermediate Neglect of Differential Overlap). MSINDO is a semi-empirical method based on the Hartree-Fock theory. It has been parametrized for the elements H, Li-F, Na-Cl, K-Br and it has shown good accuracy for molecules, comparable to more sophisticated approaches like B3LYP/6-31G. MSINDO has also been applied to bulk and surface studies of many oxides[33]. Because of this background, the choice of a semi-empirical method like MSINDO is a compromise between the quality of the model and the computational effort invested as CPU time for the calculations on the large systems of this work.

For the description of solids and surfaces, MSINDO employs the Cyclic Cluster Model (CCM)[20] which consists of a molecular cluster with periodic boundary conditions. Briefly, the basic concept of this model is summarized as follows.

As an example one can take an infinite one-dimensional chain:

∞---A-B-A-B-A-B---∞

If one starts from the model $(AB)_3$ then the molecular cluster is:

A_1-B_1-A_2-B_2-A_3-B_3

Periodic boundary conditions are introduced by translating the entire cluster by means of the translational vector a to the right and to the left once:

$$A_1\text{-}B_1\text{-}A_2\text{-}B_2\text{-}A_3\text{-}B_3\text{-}\underbrace{A_1\text{-}B_1\text{-}A_2\text{-}B_2\text{-}A_3\text{-}B_3}_{\xrightarrow{a}}\text{-}A_1\text{-}B_1\text{-}A_2\text{-}B_2\text{-}A_3\text{-}B_3$$

And then defining a limited interaction region $\pm\ |a|/2$, around each atom which corresponds to a one-dimensional Wigner-Seitz Cell:

$$A_1\text{-}B_1\text{-}A_2\text{-}B_2\text{-}A_3\text{-}B_3\ \boxed{A_1\text{-}B_1\text{-}A_2\text{-}B_2\text{-}A_3\text{-}B_3}\ A_1\text{-}B_1\text{-}A_2\text{-}B_2\text{-}A_3\text{-}B_3$$

Wigner-Seitz Cell (1D)

The two advantages of this approach are that the local symmetry is conserved, since every atom of the same kind has the same environment and that there are no dangling bonds as in molecular cluster models. One can additionally take into account long range Coulomb interactions by the Ewald summation technique.

First, a convergence test of several bulk properties with respect to model size has been carried out. For this reason, clusters of different sizes were optimized and their results compared. The clusters employed expose the crystallographic face (100). This surface was selected since it was found to be the most abundant in the nanocrystals using HRTEM[21]. Moreover, this face appears to be, from periodic Hartree – Fock and density-functional theory calculations, among the thermodynamically most stable ones like the (101) and the (001) faces[22, 23].

Following a careful inspection of the results of the employed clusters, it was decided to perform the calculations on the system, selecting the cluster 3 x 2 x 4 (the dimensions of the cluster correspond to the number of repeated units translated in the respective directions X, Y and Z), as a compromise between calculation time and quality of the results. This cluster is the smallest that shows reasonable convergence of calculated bulk properties such as lattice parameters and binding energy. Therefore, for the purpose of this study this model appears to be the optimal choice (see Figure 1).

The FTIR spectra simulation

The FTIR spectra have been simulated employing the Born-Oppenheimer Molecular Dynamics method[24]. The Newton equations of motion for the atoms are solved numerically using velocity Verlet algorithm.

To generate the vibrational spectra the velocity autocorrelation function has been employed in combination with a Fourier transformation.
Again, as a compromise between quality and computational time, the selected conditions were: 2000 time steps of 1 fs of equilibration at a temperature of 300 K using the Nosé Hoover chain thermostat. A convergence test with respect to the number of equilibration steps was performed first. The spectral changes concerning intensities and frequencies are considerably small even after a large number of additional steps. For the qualitative analysis performed in this work the smallest number of steps was acceptable and was therefore used in all further simulations.

RESULTS AND DISCUSSION

Monitoring the surface and the supernatant solution

Figure 2A shows the time evolution of the FTIR spectra of adsorbed oxalic acid on anatase in the dark. The arrows above the peaks indicate that their intensities increase during the acquisition time of 2 hours until the system reaches its equilibrium state. This moment is characterized by the fact that no further spectral changes are recorded. The last spectrum corresponds to this steady state and is shown in red.

Figure 2B shows the time evolution of the FTIR spectra of the above mentioned system under steady state UV(A) illumination. The spectrum that corresponds to the equilibrium situation of Figure 2A (in red) has now been drawn in blue while, the last spectrum after 8 hours of illumination is shown in red. The arrows indicate that, counting from the highest wavenumbers, the intensities of the second, the third and the fifth peak increase, while only the intensity of the first peak decreases and the intensity of the fourth peak remains approximately constant.

Measurements of the concentration of oxalic acid, denoted as [Ox], in the supernatant solution above the thin anatase layer, have been carried out by means of high performance liquid chromatography. The sub-indeces initial, eq and illum indicate the corresponding oxalic acid concentration: initially, after equilibration and after illumination, respectively. Table 1 summarizes the HPLC results.

Putting both these informations together, the adsorbed and dissolved oxalic acid concentration, $Ox_{ads,\,eq}$ can be attributed to the integrated area of the spectrum in the equilibrated state ($Area_{eq}$), thus it is possible to obtain by a direct relation the number of moles of oxalic acid adsorbed after illumination ($Ox_{ads,\,illum}$) from the integrated area of the last spectrum measured under this condition ($Area_{illum}$).

Taking into account a constant value corresponding to a certain number of moles of oxalic acid adsorbed within the rest of the system (remote surfaces, for example the internal walls of the transporting tubes, the measuring cell, etc), the quantity of oxalic acid adsorbed after equilibration, $Ox_{ads,\,eq}$ can be estimated (from indirect experimental values[18]) and the calculation yields the number of moles oxalic acid for anatase:

$$Ox_{illum} + Ox_{ads,\,illum} < Ox_{initial}$$

Therefore, there must be some (photo-catalytic) degradation in the system.

In this manner, from the FTIR results presented so far, several questions arise immediately:

Which is (are) the adsorption structure(s)?
What is the reason for the increasing surface concentration after illumination?
What are the intermediates formed during the photodegradation on anatase? And do they adsorb to the TiO_2 surface?

As an attempt to answer the above mentioned questions, quantum chemical calculations have been performed.

Calculated adsorption structures

The first step was to perform calculations of the water adsorption on the TiO_2 surface, since all systems are in equilibrium with aqueous solutions. Our results showed that at anatase (100) dissociated water is thermodynamically more stable than the molecular form[25] for full monolayer coverage where the number of adsorbed water molecules is identical to the number of surface Ti atoms. Therefore, this has been used in the subsequent models to represent the water adsorption.

The second step was to carry out a complete and systematic study concerning the adsorption of oxalic acid. This has been done starting from different initial configurations, considering those that have been suggested in the literature, e. g., in our previous investigations[12, 14, 15] or in other studies[16, 18] and others which have been created in order to complete the principal possibilities.

The basic criterion for stability is the calculated adsorption energy which is the difference between the total energy of the molecule-surface complex and the separate systems, the gas-phase molecules and the hydrated surface. Figure 3 shows the most stable complexes calculated here. They have been denoted as A and B, respectively. Their similarities consist not only in the loss of the co-planarity of the two carbonyl groups, but also in their bi-dentate mode of adsorption, in agreement with periodic Hartree – Fock calculations[26]. The two main differences between them are first, the spatial location of the carbon – carbon bond with respect to the TiO_2 surface: parallel in case of structure A and perpendicular in case of structure B; and second, the degree of protonation, with the structure A being totally deprotonated and the structure B being monoprotonated. From the point of view of the experimentally employed pH (3.70), both structures can coexist.

Possible intermediates

Experimental evidence for the possible formation of intermediate compounds during the photo-degradation of the oxalic acid has been searched. For this purpose adsorption studies of formic acid and carbonate on TiO_2 in its anatase crystalline phase have been carried out. These two compounds represent possible intermediates during the photocatalytic degradation of oxalic following the cleavage the sigma bond between the two carbon atoms. Thus, the remaining adsorbed structure on the TiO_2 surface should yield only two possible intermediates:

formic acid by means of the transfer of one hydrogen atom to the carbon atom (resembling the most stable conformation on TiO_2[27]) and the loss of a CO_2 molecule, or
carbonate, if the remaining adsorbed atoms combine with a third oxygen atom from the TiO_2 matrix, as reported in the literature for the mechanism of adsorption of carbonate[28, 29].

Figures 4A and 4B show the FTIR spectra measured on the anatase S-230 sample for adsorbed oxalic acid before and after illumination in comparison with the FTIR spectra of the equilibrated adsorbed formic acid and carbonate, respectively. It is important to remark that the adsorption of formic acid has been carried out under the same experimental conditions as employed for oxalic acid. This means that the same pH and the same concentration of acid, 2×10^{-3} $mol.dm^{-3}$ have been employed, which will be considerably higher than the quantity of formic acid expected to be formed in case of its photo-production. For the case of carbonate, the system even had to be forced in order to obtain the desired adsorption spectrum. This means that the pH had to be raised to 8.50 (much higher than pH 3.70 used in oxalic acid) and the concentration of carbonate was two orders of magnitude higher than the initial concentration of oxalic acid since carbonate adsorbs very weakly and desorbs very fast.

After inspection of these results, for both cases, these possibilities can be rouled out as realistic intermediates adsorbed on the TiO_2 surface at any relevant quantity after illumination since, neither the intensities nor the absolute positions of the frequencies match the spectral changes observed under UV illumination.

Therefore, the augmentation in the intensity of the peaks under illumination, once the system has reached equilibrium, must be due to an increased adsorption of oxalic acid. In this case the spectral shifts in the frequencies of the bands, which are minor, are most likely due to slight changes in the bulk or surface properties of the system under steady state UV(A) illumination[30].

If this assumption holds, the equilibrium state reached in the dark cannot be represented by a complete monolayer coverage of adsorbed oxalic acid, but there must rather be some "active sites potentially available" once the light arrives to the system to allow further adsorption of oxalic acid from the bulk solution. Thus, these potentially available sites will be blocked by adsorbed water molecules in the dark, hence the equilibrium state should constitute a mixed adsorption of oxalic acid and water. Once the UV(A) photons reach the system, some molecules of water are desorbed from the surface thus opening the "potentially available" active sites for the adsorption of more oxalic acid. This would justify the increase of the signal in the FTIR spectra after and during UV(A) illumination within the minor shift in their frequencies.

For this reason, assuming that the surface of the TiO_2 is fully hydrated before its contact with the oxalic acid solution, the adsorption energy of the selected most stable complexes on hydrated anatase (100) have been calculated as a simple model to account for the replacement of water to judge whether the proposed water replacement is thermodynamically possible.

Calculated adsorption energies on hydrated surfaces for the most stable complexes

Figure 5 shows schematically the reaction taken into account in the calculation of the adsorption energy for the most stable structures on anatase (100) (see also Figure 6).
The values calculated for the adsorption energies of oxalic acid on the hydrated surfaces of anatase show that the most stable structures will in principle be the most likely candidates in representing the adsorption structures on TiO_2. They are -139 and -118 $kJ.mol^{-1}$ for the species A and B, respectively.

The next step was to simulate their FTIR spectra and to compare with the experimental spectra and thus to obtain more information on the system.
Experiments vs. Theory

Figures 7A, 7B, 7C and 7D show the comparison of the FTIR spectra for the adsorption of oxalic acid obtained from the experiments and from the simulations, respectively.
The entire window of wavenumbers can be divided in two main parts: the carbonyl zone from 1800 to 1500 cm^{-1} which includes the first, second and third peak; and the molecular zone from 1500 to 1200 cm^{-1}, with C-C stretching modes and bending modes of C-C-O and C-O-O confirmed from the analysis of the simulated spectra.

With respect to the simulated spectra for the structures A (red) and B (blue) two main differences are observed. Concerning the carbonyl zone, the position of the main peak around 1740 cm^{-1} is shifted towards longer wavenumbers for structure A and towards shorter wavenumbers for structure B. Tentatively these two main peaks are therefore assigned to the first and the second peak from the experimental spectrum of the adsorbed oxalic acid on anatase. Another main difference between the simulated FTIR spectra for the structures A and B appears in the molecular zone, where only structure B presents two pronounced peaks at 1429 and 1270 cm^{-1} that match the obtained experimentally ones very well.

Assuming only the presence of the complexes A and B, in principle, the situation of the oxalic acid adsorption in the dark can be completely explained. The equilibration on the surface leads to a mixture of adsorbed molecules of water and oxalic acid (as justified before), with the acid adsorbing in the two different structures A and B that are also in equilibrium with each other. Thus, all experimental signals are a result of the linear combination of the signals produced by these two structures.

The situation under illumination is not so easily explained. Apparently, the concentration of the surface adsorbed species A increases upon illumination while that of species B decreases. This latter decrease can be explained by the photodegradation of this species which might be more readily attacked than species A because the sigma bond between the two carbon atoms is more exposed to any free radical attack. It is thus possible to assume that the adsorption sites loosing species B due to its photocatalytic degradation subsequently will be available as adsorption sites forming the structure A. Alternatively; adsorbed water molecules can be lost during illumination by photodesorption enabling the further formation of species A. Similar mechanisms have been proposed for the photo-desorption of water in illuminated TiO_2 suspensions[31, 32] employing very different experimental methods, thus supporting this new hypothesis.

SUMMARY

Figure 8 shows a summary of the main processes that have been investigated in this work.
The initially hydrated TiO_2 surface upon contact with the oxalic acid solution in the dark reaches an equilibrium that is a mixture of adsorbed water and oxalic acid molecules. The oxalic acid molecules adopt the conformation of the species A and B which are in equilibrium with each other since they are quasi iso-energetic. Upon UV illumination of the system, several changes occur at the surface. Some molecules of water are depleted from the surface leaving free active sites for adsorption of more oxalic acid molecules, which under these non-equilibrium conditions apparently prefer the conformation of the species A. The surface is thus enriched in A and

relatively depleted in B. In parallel, some photocatalytic degradation proceeds for the oxalic acid adsorbed probably primarily by the attack on the more exposed species B, thus enabling further shifts of the adsorption equilibrium in favor of complex A.

Acknowledgement
We are grateful for the support by Millennium Inorganic Chemicals (MIC), a Lyondell Chemical Company and for the continued interaction with the responsible MIC scientists: Claire Bygott and Rob McIntyre. We like to thank the University of Hannover, Prof. Buhl and Dr. Gesing for the XRD, Dr. Dringenberg for the BET measurements and Dr. Feldhoff for the HRTEM pictures.

REFERENCES

[1]Hoffmann M., Martin S., Choi W. and Bahnemann D. *Chem. Rev.* 95 (1995) 69
[2]Legrini O., Oliveros E. and Braun A.M. *Chem. Rev.* 93 (1993) 671
[3]Bahnemann D., Cunningham J., Fox M.A., Pelizetti E., Pichat P. and Serpone N., *Photocatalytic treatment of waters*, in: G.Helz, R. Zepp and D. Crosby (Eds.), *Aquatic and Surface Photochemistry*, CRC Press, 1994, 261 pp.
[4]Goswami D.Y., *Engineering of the solar photocatalytic detoxification and disinfection processes*, in: K.W. Boer (Ed.), *Advances in Solar Energy*, vol. 10, American Solar Energy Society, Boulder, CO, 1995, 165 pp.
[5]Maness P., Smolinsky S. and Jacoby W. *Appl. Environ. Microbiol.* 65 (1999) 4094
[6]Paz Y., Luo Z., Rabenberg L. and Heller A. *J. Mater. Res.* 10 (1995) 284
[7]Poulius I., Spathis P. and Tsoumparis P. *J. Environ. Sci. Health* 34 (1999) 1455
[8]Cai R., Hashimoto K., Itoh K., Kubota Y. and Fujishima A. *Bull. Chem. Soc. Jpn.* 64 (1991) 1268
[9]Fujishima A., Cai R., Hashimoto K., Sakai H. and Kubota Y. *Trace Met. Environ.* 3 (1993) 193
[10]Sakai H., Baba R., Hashimoto K., Kubota Y. and Fujishima A. *Chem. Lett.* (1995) 185
[11]Heller A., *Acc. Chem Res.* 28 (1995) 503
[12]Mendive C., Bahnemann D. and Blesa M. *Catalysis Today* 101 (2005) 237
[13]Araujo P., Mendive C., García Rodenas L., Morando P., Regazzoni A., Blesa M. A. and Bahnemann D. *Colloids and Surfaces A: Physicochem. Eng. Aspects* 265 (2005) 73
[14]Mendive C., Weisz A. and Blesa M., 11th International Conference on Surface and Colloid Science, 2003, Foz do Iguacu, Brazil
[15]Mendive C., Benitez Lorca F., Bahnemann D., and Blesa M., 3rd European Meeting on Solar Chemistry and Photocatalysis: Environmental Applications (SPEA3), 2004, Barcelona, Spain
[16]Weisz A., García Rodenas L., Morando P., Regazzoni A. and Blesa M., *Catalysis Today*, 76 (2002) 103
[17]Weisz A., Regazzoni A. and Blesa M., *Solid State Ionics*, 143 (2001) 125
[18]Hug S. J. and Sulzberger B. *Langmuir* 10 (1994) 3587
[19]Hug S. J., *J. Colloid Interface Sci.*, 188 (1997) 415
[20]Bredow T., Geudtner G., Jug K. *J. Comput. Chem.* 22 (2001) 861
[21]Feldhof A. Personal communication
[22]Lazzeri M., Vittadini A. and Selloni A. *Phy. Rev. B* 63 (2001) 155409/1
[23]Lazzeri M., Vittadini A. and Selloni A. *Phy. Rev. B* 65 (2002) 119901/1
[24]Nair N.N., Bredow T. and Jug K. *J. Comput. Chem.* vol 25, 10 (2005) 1255
[25]Barnard A., Zapol P. and Curtiss L. *Surf.Scie.* 582 (2005) 173
[26]Fahmi A., Minot C., Fourré .P and Nortier P. *Surf. Sci.* 343 (1995) 261

[27]Rotzinger F., Kesselman-Truttmann J., Hug S., Shklover V. and Grätzel M. *J. Phys. Chem. B* 108 (2004) 5004
[28]Onishi H., Aruga T., Egawa C., Iwasawa Y. *Sur. Sci.* 199 (1988) 54
[29]Onishi H., Aruga T., Egawa C., Iwasawa Y. *J. Chem. Soc., Faraday Trans.* 85 (1989) 2597
[30]Szczepankiewicz S., Colussi A. and Hoffmann M. *J. Phys. Chem. B* 104 (2000) 9842
[31]Wang C., Pagel R., Bahnemann D. and Dohrmann J. *J. Phys. Chem. B* 108 (2004) 14082
[32]Sakai N., Fujishima A., Watanabe T. and Hashimoto K. *J. Phys. Chem. B* 107 (2003) 1028
[33]Bredow T., Jug K. *Theoret. Chem. Acc.* 113 (2005) 1 -14

Table 1. Oxalic acid concentrations in the supernatant solutions above the anatase and the rutile particle layers.

Oxalic acid in solution [x 10^{-3} mol.dm^{-3}]	**TiO_2 – Anatase**
$[Ox]_{initial}$	1.94
$[Ox]_{eq}$	1.81
$[Ox]_{illum}$	1.56

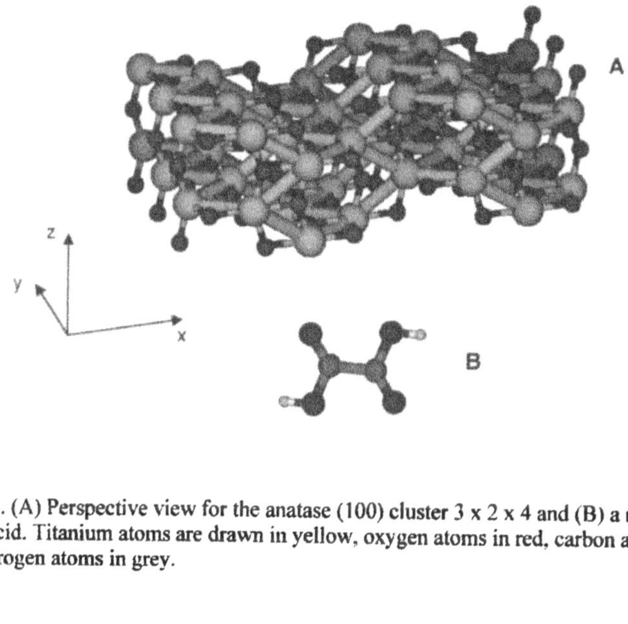

Figure 1. (A) Perspective view for the anatase (100) cluster 3 x 2 x 4 and (B) a molecule of oxalic acid. Titanium atoms are drawn in yellow, oxygen atoms in red, carbon atoms in green and hydrogen atoms in grey.

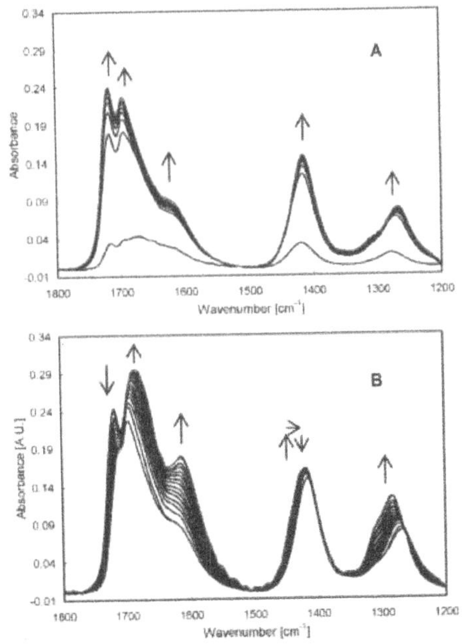

Figure 2. FTIR spectra of adsorbed oxalic acid on anatase. (A) Evolution of the system oxalic acid on TiO_2 (S-230) in time in the dark, the last spectrum (red) corresponds to the equilibrated state; and (B) evolution of the same system under UV illumination, in blue, the equilibrated state reached in the dark, and in red, the last spectrum acquired under illumination.

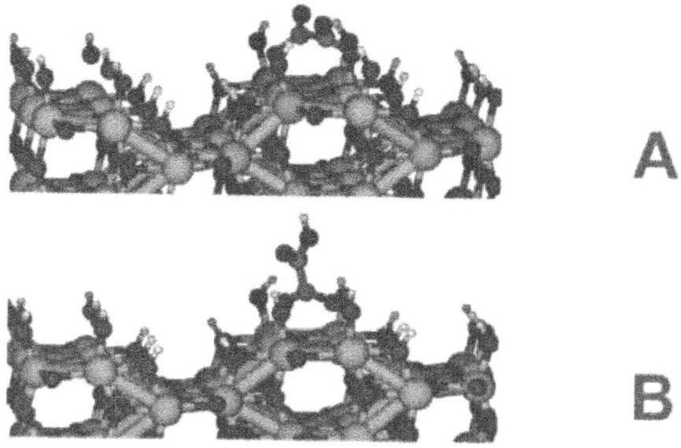

Figure 3. Most stable adsorbed oxalic acid on hydrated anatase (100). The complexes correspond to bi-dentate structures named A and B.

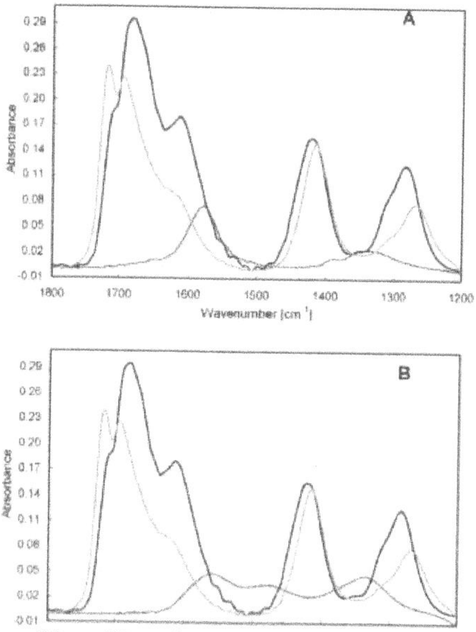

Figure 4. (A) FTIR spectra on anatase S-230 of (green) equilibrated adsorbed oxalic acid, (black) last stage of illumination of the same system and (pink) equilibrated adsorbed formic acid. (B) FTIR spectra on anatase S-230 of (green) equilibrated adsorbed oxalic acid, (black) last stage of illumination of the same system and (pink) equilibrated adsorbed carbonate.

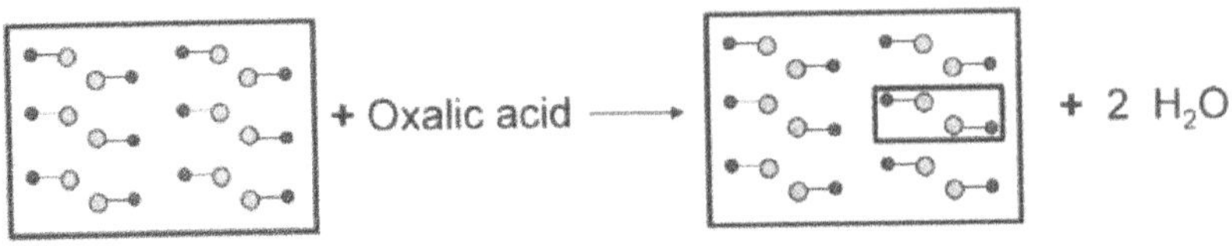

Figure 5. Cluster of anatase (100). (a) Side view. (b) Top view. (c) Schematic representation of the top view, the active sites are identified in yellow for Ti and in red for O.

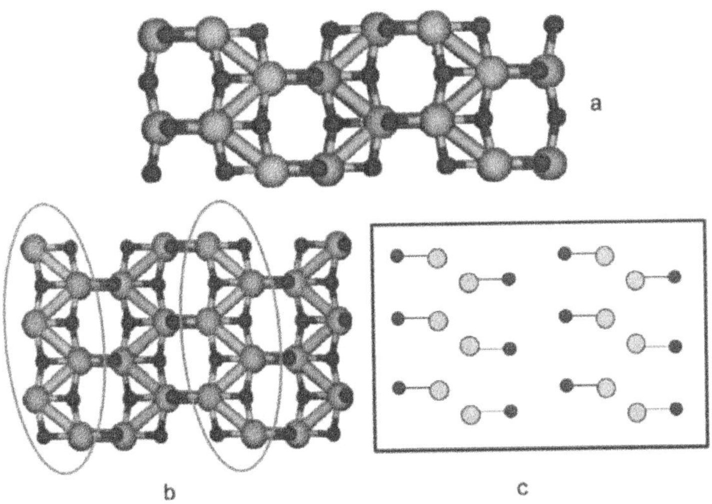

Figure 6. Adsorption reaction for the formation of the most stable structures on hydrated anatase (100), replacement of two molecules of water.

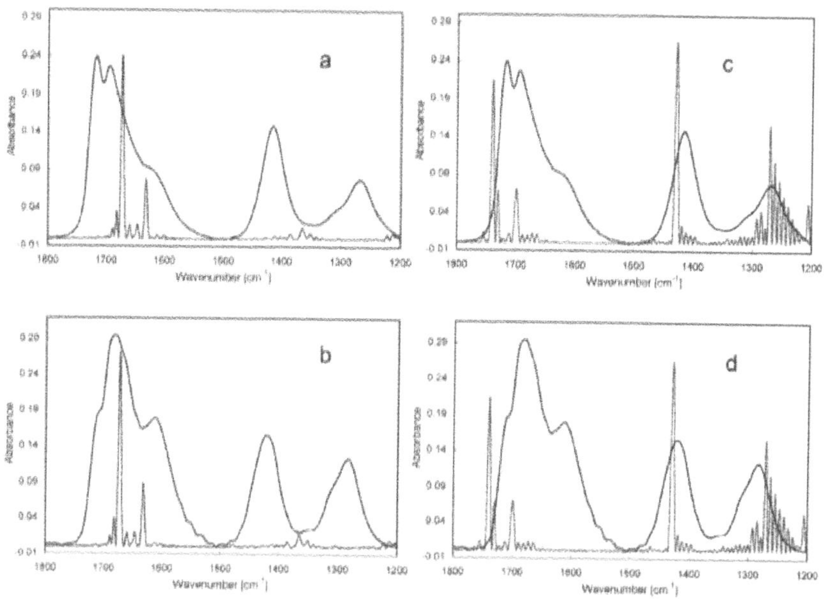

Figure 7. Experimental FTIR spectra (in black) and simulated FTIR spectra (in red or blue) of adsorbed oxalic acid on anatase. (A) black – FTIR spectrum of equilibrated adsorbed oxalic acid in the dark on anatase (S-230), red – simulated FTIR spectrum of the optimized structure A on anatase (100). (B) black – FTIR spectrum of the last stage of UV illumination of adsorbed oxalic acid on anatase (S-230), red – simulated FTIR spectrum of the optimized structure A on anatase (100). (C) black – FTIR spectrum of equilibrated adsorbed oxalic acid in the dark on anatase (S-230), blue – simulated FTIR spectrum of the optimized structure B on anatase (100). (D) black – FTIR spectrum of the last stage of UV illumination of adsorbed oxalic acid on anatase (S-230), blue – simulated FTIR spectrum of the optimized structure B on anatase (100).

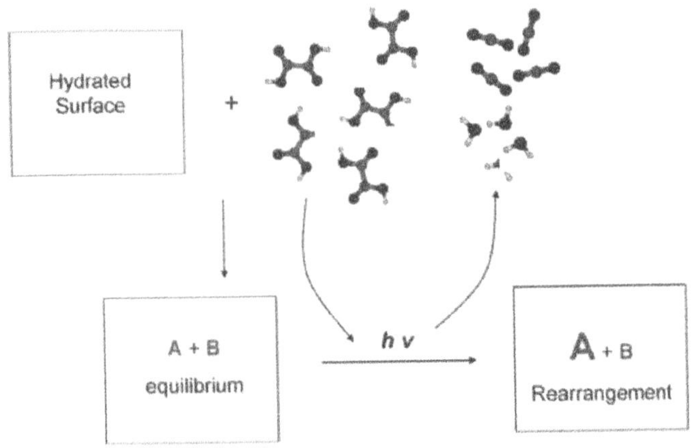

Figure 8. Summary of the processes.

Eco-Materials Processing and Design
Edited by Koji Watari and Soo-Wohn Lee

EFFECTS OF CONSTITUENTS IN PHOTO/BIOCATALYTIC HYDROGEN PRODUCTION SYSTEM USING EXPERIMENTAL DESIGN TOOL

Hyunku Joo, Yongho Jang
Hydrogen & Fuel Cell Research Dept., Korea Institute of Energy Research
71-2 Jang-dong, Yusong-ku
Taejeon, 305-343 Rep. of Korea

Sangbong Lee, Yong-gun Shul
Chemical Engr. Dept. & Center for Clean Technology, Yonsei University
134 Shinchon-dong, Seodaemun-ku
Seoul, 120-752 Rep. of Korea

ABSTRACT

This study photoelectrochemically and photocatalytically examined the effects of constituents in photo/biocatalytic hydrogen production system, being eventually composed of photoanode and cathode, just after that preliminary mixed slurry system had been investigated to see the insights of the hybrid system.

Experiments with the mixed slurry system revealed the following four cases; direct inter-phase electron transfer from photocatalyst to enzyme in the absence of electron relay was a rate-determining step, enzyme was deactivated by irradiated light, physical properties of photocatalysts, obtained by a couple of preparation methods, seldom affected the production rate, and each material needs different reaction conditions such as pH and temperature. These results led the study to the system where the materials were separated into photoanode and cathode and each material has been studied individually and simultaneously to get the possible optimum condition: for the former photoanodic current and photovoltage were measured for each prepared photocatalysts and for the latter change in the amount of evolved hydrogen was checked according to external bias. Then both reaction parts were connected through Nafion membrane and salt bridge to perform light-induced hydrogen production for the hybrid system. No synergistic effect of the system happened with Nafion, but promising result was obtained with salt bridge. Detailed investigations are under way.

INTRODUCTION

Environmental remediation using photocatalysts has been a major focus in many countries over the last decade. Also, numerous studies have been focused on hydrogen from water as a clean energy resource since the invention of photochemical water splitting over TiO_2 electrode. However, the photocatalytic process has been simultaneously criticized as being uneconomical compared to other oxidative treatment systems and hydrogen production systems up to date due to its inherent low efficiency and limitations resulting from the necessities for an appropriate light source and immobilization, which may increase the overall energy costs[1]. Therefore, there has been a great deal of interest in tackling the aforementioned key issues to economically commercialize photocatalytic systems, even though the effort has been taken as a long-termed plan.

As far as photo-production of hydrogen is concerned, one of the IEA hydrogen programs (known as the Hydrogen Implementing Agreement (HIA)), Annex-14, focused on the development of materials and systems for the photoelectrochemical (PEC) production of hydrogen[2]. On the other hand, in Korea since 2003 the 21st Century Frontier R&D Program by the Hydrogen R&D Research Center has brought together a relatively broad range of topics to address challenges with a long term global view. Using a PEC composed of a TiO_2 single crystal as photoanode and Pt as cathode involving a chemical bias imposed by pH difference between the electrodes, Fujishima and Honda claimed that the total efficiency of the solar energy conversion to the hydrogen is 0.4%. Thereafter, the search for other oxide materials that do not require external bias and exhibit the higher light conversion has been studied by many researchers, resulting in that TiO_2 and TiO_2-based materials were identified so far as the best candidate for photoanode[3]. Hence, an approach, photo/biocatalytic system, other than development of photo-active new or modified materials has been thought for consideration. This approach, being enzyme as a main consideration, has been studied by many researchers since in the middle of 80's, while the present study has been performed from the photo catalytic point of view. The system naturally uses the intrinsic nature of proton reduction ability of hydrogenase enzyme by help of photoanode being electron donors under solar irradiation.

In in-vitro biological system for hydrogen production, complex electron transport systems exist such as photosystem (PS) II and I, and ferrodoxin (FD). This electron transport system can be simply theoretically replaced with visible-light sensitized photocatalyst. Thus, the photocatalyst sensitized by solar irradiation produces electrons and holes (electron vacancies) which then can be separated to reduce protons and to oxidize electron donor (Fig. 1). For the photocatalyst particles to be activated by visible light, nitrogen-substituted TiO_2 (TiON) has been studied, either by sputtering or by hydrolyzing an aqueous $Ti(SO_4)_2$ solution with a NH_3 solution[4-5]. Moreover, the efficient charge separation is not related to absorption of light with the wavelength shorter than that of bandgap energy, but depends on the physical properties of the samples such as the shape, distribution pores, crystalline phase, crystallinity, and the preparation and the post-deposition treatments, even though they must be considered with substrate at the same time. Hence, the effect of physical properties tried to be considered. In addition, though used enzyme from Pyrococcus furiosus (Pfu), first isolated by Bryant and Adams[6], was known to be remarkably resistant to inactivation by temperature and chemical agents, the Pfu was deactivated by irradiated light and oxygen and needed different pH and temperature. For this reason, photoelectrochemical approach with separated reaction part has been performed. Factorial design has been known as a methodology for applying statistics to experimentation to see how responses (output variables) change and interact at different variable settings. This tool has a versatility and the use of it, especially fractional factorial design (FFD), can save our time and a little of cost. Therefore, this method was occasionally applied when large amount of variables existed and needed to be screened by the level of significance. In this article details were concentrated on photoelectrochemical analyses with photoanode and cathode individually and together, showing summarized results from the preliminary experiments.

EXPERIMENTAL

Visible light-activated nitrogen-substituted titanium oxide ('TiON') was simply prepared by a couple of methods (sol-gel and hydrothermal). P25 TiO_2 (Degussa, FRG), UV100 (Hombikat, FRG) and NT22 TiO_2 (NANO Co., Ltd. Korea) were used as a raw material and a reference when necessary. For hydrothermal TiON ('HT-TiON'), anatase TiO_2 powders were placed into a teflon-lined autoclave with a 50ml capacity. The autoclave was then filled with a aqueous solution of NaOH with a certain molar concentration up to 80% of the total volume of the autoclave and was maintained under the desirable conditions without shaking or stirring during the heating process. Washing and drying was then performed. To narrow the conditions the one-third fractional factorial experiment involving 9 FLC was used to reduce the size of the experiment: a full factorial experiment a 3^3=27 factor-level combination (FLC) occurs when there are three factors and three levels for each factor. The selected factors consist of the NaOH concentration (levels of 1, 3 and 5M), temperature (levels of 175, 220 and 225 °C) and time (levels of 12, 24 and 36 hr). As a preliminary experiment, raw TiO_2 for TiON was prepared using titanium tetraisopropoxide (TTIP), isopropyl alcohol (IPA), acetic acid and water for comparison. Finally, the substitution of nitrogen into the hydrothermally treated TiO_2 was performed by mixing TiO_2 powder with a NH_3 solution followed by drying (Fig. 2). TiON prepared at low temperature ('LT-TiON') was prepared using low temperature synthesis with triethylamine (TEA), resulting in anatase crystallite phase and ca. 170 m^2/g of surface area.

Purified hydrogenase (Pyrococcus furiosus, Pfu) was purchased from Prof. Adams at the University of Georgia. Activity of Pfu was stable for at least 48 hr after it was out of a freezer. And activity assay of Pfu (21834 unit/ml) was much higher than that of Clostridium butyricum (1442 unit/ml) and Thiocapsa roseopersicine (1704 unit/ml) determined at Tris-HCl (50mM, pH 8.5, 50 °C, absorbance at 570 nm). Specific activity (unit/mg, Bio-rad protein assay, absorbance at 750 nm) of Pfu was two to three times higher than that of the latter hydrogenases. Activity was lower in phosphaste, but similar in EPPS buffer. Methyl viologen (MV), tris(hydroxymethyl)aminomethane (Tris), phosphate, MES, sodium formate, and sodium dithionite (Na-D) were obtained from Sigma and used as received.

Photoelectrochemical experiment was conducted in a two-compartment (anodic and cathodic one connected via a Nafion membrane or salt bridge) eventually, but started in a compartment separately. Anodic compartment has a volume of 60 ml and was a cylindrical-shaped cell. Photocatalyst films were prepared by casting the photocatalyst sol mixed with polyethylene glycol (PEG from Fluka, MW 20,000) as a binder onto indium-tin oxide (ITO) conducting substrate, and subsequent heat treatment at 450 °C for 30 min under air (ca 10mg coated in 1×1 cm^2). The measurement of cyclovoltammograms and photovoltage was performed with a potentiostat (EG&G, Model 273A) with a SUS316 electrode and Ag/AgCl (saturated in KCl) as a counter and reference electrode, respectively, in the anodic compartment (an aqueous solution consisting of 0.1M Na_2S and 0.02M Na_2SO_3). Cathodic experiments were performed in a water-jacketed cylindrical-shaped cell with a buffer EPPS (50 mM, pH 8) sealed with silicone rubber gasket.

The light sources used were UV-A lamps (Sankyo Denki, Japan), halogen lamp, xenon lamp (1000W output max., Oriel, USA. w/ water filter) and LED (green and blue peaked at 520 nm and 464 nm, respectively.). Isopropyl alcohol (IPA) was a probe compound for measuring the activity. The chemicals (IPA, acetone and CO_2) were analyzed by gas chromatography with a flame ionization detector (GC/FID, HP 5890) and a GC/FID with a methanizer (HP6890). Hydrogen production was analyzed by gas chromatograph with TCD (thermal conductivity detector at 160 °C). The used column in the system was molecular sieve 5A. Typical oven temperature was maintained at 40 °C. Moisture trap (silica-gel) was placed just before GC auto injection valve to prevent damage by water. The structure, morphology and the surface area were investigated using XRD (Rigaku, DMAX/2000-Ultima Plus), SEM (Topcon SM-720), and BET.

RESULTS AND DISCUSSIONS

Initially, a rough hydrothermal experiment with NH_4OH and the prepared raw TiO_2 (189 m^2/g) revealed that the sample turned a yellow color and absorbed light with wavelength < ca. 520 nm. In addition, using aqueous NaOH (10 M) and a 12hr reaction time under varying reaction temperatures ranging from 100 ~ 200 °C, five samples (with 25 °C interval) exhibited the beta phase TiO_2 (β-TiO_2; Na_2TiO_3, $Na_2Ti_3O_7$) from XRD analysis with a maximum specific surface area of 246 m^2/g at 175 °C. The addition of silica made the particle size more uniformly distributed (a fibrous shape was identified by SEM[7]). The following properties, specific surface area, pore radius and pore volume, were changed with respect to the washing procedure used, showing a linear relationship between the specific surface area and pore radius (Table 1). This suggests that washing and its procedure are essential for removing impurities and controlling the shape.

Systematic approach for preparing the photo-active TiO_2 was made using a fractional factorial design (FFD), a tool for experimental setup[8] (Table 2). From an analysis of the variance (ANOVA), reaction temperature had 99% level of significance on the specific surface area. The NaOH concentration of was critical to the phase formed. The anatase phase was formed with 1M of NaOH, the β phase was formed with 5M of NaOH, and for 3M of NaOH the anatase was formed below 225 °C and β phase above 225 °C. The specific surface area increased from 65 m^2/g to 144 m^2/g for 5M NaOH, 225 °C, 12 hr. A fibrous shape of TiO_2 was revealed at concentrations > 3M of NaOH. Detailed investigations based on the 3M NaOH, 175 °C and 12hr were examined in terms of the phase and shape, varying the NaOH concentration (3.5, 4, 4.5 M) first and then followed by altering the reaction temperature and time. The resulting optimum condition was found to be 4M of NaOH, 200 °C and 12 hr (103 m^2/g). A summary of the physical properties of the selected samples is shown in Table 3, and their photocatalytic activity in degrading IPA using UV-A irradiation was investigated where P25 showed the fastest reaction rate, followed by NT22. The optimized sample (4M200T12-8/2w) showed a better activity than the other samples prepared, but was less active than the two commercialized TiO_2 samples under UV-A irradiation. The observed change in surface area is attributable to the crystallization of walls separating mesopores. This tendency subsequently caused a change in mean pore diameter and pore volume of the bulk materials. Consequently, the loss in surface area caused an increase in pore diameter and a decrease in pore volume[9]. The optimized sample (4M200T12-8/2w) was further analyzed by DSC to identify the heat-treatment temperature to substitute nitrogen, where a distinguishable exothermic band appeared near 430 °C. This exothermic reaction was attributed to be the oxidation reaction of NH_3 or NH_2 with the oxygen released from the amorphous grain boundaries by forming oxygen-deficient sites[5]. This band was not significant due to the small amount of the amorphous portion in the obtained samples by the hydrothermal method. This fact needs to be further studied in terms of the ease and the extent of nitrogen-substitution. As a consequence, NH_3 treatment, followed by heat-treatment at 450 °C produced yellow powders (HT-TiON, Fig. 3). A commercially available blue LED (peaked at 464 nm) was used to check the photocatalytic activity of the obtained HT-TiON (Fig. 4). Acetone evolution as a function of blue LED irradiation time was remarkable for the HT-TiON sample, but no acetone was detected for the NT22 sample (P25 also revealed no acetone production in other study by authors, data not shown).

To obtain the direct insights for photo/biocatalytic hydrogen production system, two sets of one-third fractional factorial experiment involving 9 FLC was used in order to reduce the size of the experiment: a full factorial experiment 3^3=27 factor-level combination (FLC). The factors included species of buffer solution, the amount of hydrogenase (Pfu), electron donor, reaction temperature and electron mediator. The correlation of electron mediator and Pfu was also studied by a half of 2^4 designs (8 FLCs). For a fixed hydrogenase (Pfu), four factors $CdCl_2$, Na-dithionite, sodium formate and MV was tested in terms of presence (two levels) in a designed manner. The results from the ANOVA showed that the amount of Pfu and Pfu+species of buffer were significant at the 0.25 significance level for the H_2 production. And Pfu+pH of Tris was significant at the 0.1 significance level and each of factor A and B were significant at the 0.025 significance level for the H_2 production. The two-factor interactions showed significant effect on the response variables (Table 4). It can be generally said that mentioned photo/biocatalytic system is described as a possible candidate for hydrogen production technique with solar irradiation. Rough photonic efficiency was calculated as follows, eq. (1) & (2).

$$\phi = \frac{K_{H2}}{K_{irr}} \tag{1}$$

where K_{H2} denotes the amount of produced H_2 in mol/s and K_{irr} irradiated light in mol (photon)/s.

$$K_{irr} = \int \frac{I}{U_\lambda} d\lambda = \int \frac{I \times \lambda}{hcN_A} d\lambda \tag{2}$$

Here, I represents measured light intensity [W], h Plank constant [$6.6260755 \times 10^{-34}$ J·s], c speed of light in vacuum [3×10^8 m/s], N_A Avogadro number [6.0221367×10^{23} mol^{-1}] and λ representative wavelength of irradiated light [386 nm]. Measured total intensity I was ca. 700 μW/cm^2 (in the range of 310~386 nm of used halogen lamp. This measured value is about one fifth of solar UV irradiation on earth, 3~4 mW/cm^2). Hence, resulted K_{irr} turned out to be 20.3×10^{-10} mol photon/s. According to a reference, solar irradiation is 1.5×10^{-6} mol/s per 1.4245×10^{-2} m^2 [10], which means 105×10^{-10} mol photon/s per cm^2. This is also five times higher than measured intensity I. As a result, with 20mg of photocatalyst approximately 5.6×10^{-10} mol H_2/s produced. Therefore, rough photonic efficiency was close to 30%. This value could be reasonably taken form the fact that intensity of the irradiated light is low and the highest H_2 evolution rate with photo/biocatalytic system was found to be 30×10^3 μmol/(hr·g)[11], which is a few hundred times higher than that with typical TiO_2-based photocatalytic systems; a couple of hundreds μmol/(hr·g). Figure 5 shows the schematic view of summarized reaction mechanism where electrons movement from photocatalyst to relay chemical was identified as a rate determining step. As a preliminary result, factors to be considered for the next progress were selected as follows. One is to separate photocatalyst and hydrogenase to overcome barriers such as instability of hydrogenase to photon and back reaction of hydrogen to water. Second is from system's point of view either to immobilize photocatalyst and hydrogenase or to connect both materials via electron relay, which consequently prevents deactivation and enhances durability.

Two compartments reactor was fabricated to conduct optimized photo/biocatalytic hydrogen production. Before experiments for hydrogen production have been conducted, photoanodic and cathodic compartments were studied individually. When a semiconductor is in contact with an electrolyte, two areas are brought together; the solid-state physics and the electrochemistry. Since these two areas use two different energy scales where the former is based on the atomic and molecular orbital theory using the energy of electron in vacuum as reference level, and the latter uses the standard reduction potential scale (on this scale the normal hydrogen electrode, NHE is by definition set to zero) to describe the redox system, a connection between two systems must be made to describe the photoelectrochemical system. This is accomplished since the Fermi level in the semiconductor and the redox potential of the electrolyte are actually the same electrochemical potential of the electron. The junction formed between an n-type semiconductor and an electrolyte when they are in contact bring about the band bending via charge transfer, and the excess charge is distributed in a space charge region. The extent of the band bending is a few tenth volts when the semiconductor is ca. 50 nm in diameter[12-13].

In the case of photoanode, nitrogen-substituted TiO_2 (TiON) powder was prepared and pasted on the conducting oxide. Then, with this photoanodic cell time coursed voltage curves (V-t, V_{oc} @ I=0), an important criterion of performance characteristics of PECs, and cyclovoltammograms (I-V curves) were drawn to obtain useful electrochemical values. The V_{oc} is electromotive force of the cell which defines the maximum possible voltage across the cell in light when no current is flowing. It also represents the energy difference between the Fermi level of semiconductor film and the reduction couple in the solution. As seen in Fig. 6, during the light-off period, the cell exhibits ca. - 0.3 V and imposition of light results in an increase of photo-voltage to the level of -1.0 V without 400 nm cutoff filter and of -0.7 ~ -0.8 with 400 nm cutoff filter. This is a quite high photo-voltage ever described in the other literatures[3, 14-15]. It is also interesting to note that the voltage drop when light-off period is substantially slower (2 ~ 3 hr) than voltage increase when the light-on period (20 ~ 60 min). It is similar to the result from Bak et al.[3], where they postulated that it was due to polarization-related phenomena. In the meantime, in Fig. 6, TiON300 and TiON400 have remarkable aspect in V-t analysis where potential drop in the presence of 400 nm-cutoff filter was not as big as that with TiON450 or P25 TiO_2 (data not shown). This showed that TiON300 and TiON400 were efficient in producing charge carriers by light in the visible range. Both TiON300 and TiON400 also revealed the highest activity for photocatalytic isopropyl alcohol degradation with blue light emitting diode (data not shown). The aforementioned band bending can be zero at a potential, called the flatband potential (E_{fb}), where no excess charge exists (no electric field and no space charge region). This flatband potential is important as it approximately gives the position of the conduction band of an n-doped semiconductor. E_{fb} can be determined by Butler equation[16] as follows. By plotting V vs. I_{ph}^2, it is possible to determine the E_{fb} when I_{ph}^2 is zero (eq. (3)).

$$V - E_{fb} \approx (I_{ph} / e\phi_o \alpha \varpi_o)^2 \tag{3}$$

When illuminated, band edges and Fermi level of an n-type semiconductor shift to more negative potential and the amount of band bending reduces. Our results showed that E_{fb} for prepared TiONs was in the range of - 0.7 ~ -0.8 V vs. SHE, so theoretically the unbiased system can produces hydrogen at pH = 9 where $E_{H^+/H2}$ is -0.53 V vs. SHE[15]. In Fig. 7 the change is the hydrogen production rate was shown with respect to applied external bias and the amount of injected chemicals for EPPS buffer in the cathode compartment. External bias (critical bias seems to be 0.5 V) was turned more necessary that the amount of chemicals in terms of hydrogen production rate.

Finally, the combination of obtained results produced a postulated energetic scheme for photo/biocatalytic hydrogen production system (Fig. 8). As shown in Fig. 8, if the anodic current starts to flow at a potential considerably more negative than the cathodic current on counter electrode, this implies that the anodic current is indicative of n-type semiconductor and the related reaction occurs in a cell spontaneously with no external bias. Besides, if the up-hill reaction with the standard Gibbs energy positive ($\Delta G>0$, e.g. water splitting) happens by solar light, called 'photoelectrosynthetic' cells, this implies that the solar energy is converted to chemical energy. Sometimes, however, depending on the relative location of redox couples in separated cells, 'photocatalytic' cells ($\Delta G<0$) was driven in the spontaneous direction with the light energy used to overcome the activation energy of the process. However, during experiments unexpected cases were brought such that the addition of MV caused detrimental effect on hydrogen evolution and Pfu seldom acted as a biocatalyst for efficient proton reduction, even though MV in cathode compartment was evidenced to be reduced turning into blue color when photoanode was irradiated. In addition, hydrogen evolution was stopped, adding MV later into the cathode compartment right after hydrogen was being produced. The strongly related characteristics between buffer and electrolyte in terms of pH, existing anions and cations in each compartment, and a kind of ion-exchange membrane may affect the results. Several efforts on these aspects have been under investigation.

CONCLUSION

For hydrogen production with solar irradiation, a relatively novel approach was started where sensitized photocatalysts give electrons to hydrogenase that reduces protons into hydrogen. The possible role of morphology and microstructure of photocatalyst in photo/biocatalytic hydrogen production system was not important, but the efficient charge carriers' separation and transfer to the enzyme was critical. Individual photoanodic and cathodic experiments showed strong feasibility of this system, while simultaneous photoelectrochemical experiments turned out unpromising. The close relation between buffer and electrolyte in terms of pH, existing anions and cations, and a kind of ion-exchange membrane may affect the results. Several efforts on these aspects have been on-going.

ACKNOWLEDGEMENTS

This research was performed for the Hydrogen Energy R&D Center, one of the 21st Century Frontier R&D Program, funded by the Ministry of Science and Technology of Korea. Authors thank Dr. Jin, Chang-Soo. at the Energy Conversion & Storage Research Center of KIER for valuable discussions on electrochemistry.

REFERENCES

[1]G. B. Raupp, A. Alexiadis, M. Hossain, and R. Changrani, *Catalysis Today*, **69**, 41-42 (2001).

[2]Photoelectrolytic production of hydrogen, IEA Hydrogen Program "Annex-14", *Final report* (Ed. Andreas Luzzi) (2004)

[3]T. Bak, J. Nowotny, M. Rekas, and C. C. Sorrell, *Int. J. Hydrogen Energy*, **27**, 19-26 (2002).

[4]R. Ashai, T. Morikawa, T. Ohwaki, K. Aoki, and Y. Taga, *SCIENCE*, **293**, 269-271 (2001).

[5]T. Ihara, M. Miyoshi, Y. Iriyama, O. Matsumoto, and S. Sugihara, *Applied Catalysis B: Environmental*, **42**, 403-409 (2003).

[6]F. O. Bryant, and M. W. W. Adams, *J. Biol. Chem.*, **264**, 5070-5079 (1989).

[7]W. Wang, O. K. Varghese, M. Paulose, C. A. Grimes, Q. Wang, and E. C. Dickey, *J. Mater. Res.*, **19(2)**, 420-421 (2004).
[8]D. C. Montgomery, "*Design and Analysis of Experiments*", John Wiley & Sons, USA (1991).
[9]T. Sreethawong, Y. Suzuki, and S. Yoshikawa, *J. of Solid State Chemistry*, **178**, 329-338 (2005).
[10]D. Ljubas, *Energy*, **30**, 1699-1710 (2005).
[11]P. Pedroni, G. M. Mura, G. Galli, C. Pratesi, L. Serbolisca, and G. Grandi, *Int. J. Hydrogen Energy*, **21(10)**, 853-858 (1996).
[12]M. Larsson, M.S. thesis for Inorganic Chemistry, Lund University, Sweden (2001).
[13]R. Memming, "*Semiconductor electrochemistry*", Wiley-VCH. Weinheim, FRG (2001).
[14]C. He, X. Li, Y. Xiong, X. Zhu, and S. Liu, *Chemosphere*, **58**, 381 -389 (2005).
[15]L. G. Arriaga, and A. M. Fernández, A. M., *Int. J. Hydrogen Energy*, **27**, 27-31 (2002).
[16]M. A. Butler, *Appl. Phys.*, 1160 (1974).

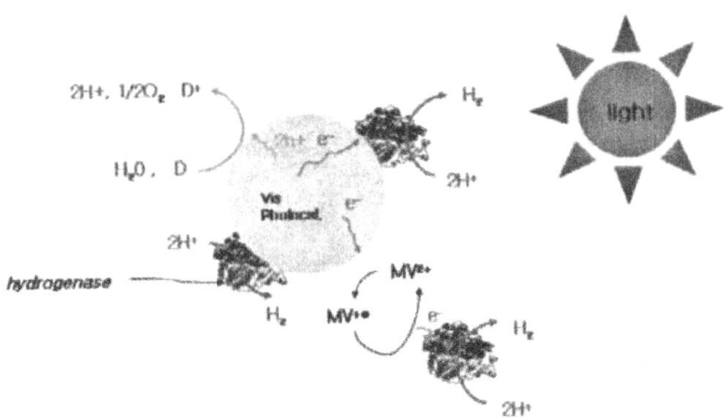

Figure 1. Schematic view of photo/biocatalytic hydrogen production system.

Table 1. Effect[1] of the washing procedure[2] on the physical properties of HT-TiO_2 from NT22 under 10M NaOH, 175 °C, 12hr.

	BET surface area (m^2/g)	Pore volume (cm^3/g)	Pore radius (nm)
NT-22	64.7427	0.22792	6.82975
W0	16.0104	0.05916	7.38978
W1	164.3932	0.33846	4.11769
W2	308.8273	0.85506	5.53749
W3	308.8627	0.88443	5.72702

[1]While impurity peaks dominate crystal phases in XRD analysis under W0, washing procedures (W1~W3) removes impurities, resulting in some characteristic crystal peaks. This can be also seen with drastically increasing surface area (or pore volume) in the Table 1.
[2]W0: without washing, W1: washing only with water, W2: washing only with HCl, W3: washing with water and HCl.

Table 2. Selected factor-level combinations (FLCs) and ANOVA analysis for the significant factor selection on surface area of TiON.

	1M NaOH			3M NaOH			5M NaOH		
	12hr	24hr	36hr	12hr	24hr	36hr	12hr	24hr	36hr
175 °C	•				•				•
200 °C			•	•				•	
225 °C		•				•	•		

Source[a] (df)	SS	MS	F_o
Total (8)			
A (2)	93.639	46.819[*]	
B (2)	1016.992	508.496	11.71[b]
C (2)	34.657	17.329[*]	
AxB (2)	132.156	66.078[*]	
Residual		43.409 (6df)	

[a] : A: mole of NHOH, B: Rxn temperature, C: Rxn time, [*] : Components involved in residuals
[b] : 0.01 level of significance.

Table 3. Physical properties of the selected samples[1].

	P25	NT22	4M200T12-(8/2w)SiO_2	4M200T12	3M225T36	5M225T12
Phase	anatase+ rutile	anatase	anatase	anatase	anatase+ β-TiO_2	β-TiO_2
XRD intensity (θ)	545 (25.18)	1392 (25.16)	688 (25.30)	528 (25.26)	454 (25.26)	262 (24.26)
1/2 width	0.36	0.42	0.43	0.43	0.46	-
BET (m^2/g)	50	64.7	127.6	102.7	52.3	114.0
Pore Vol. (cm^3/g)	-	0.228	0.448	0.365	0.189	0.566

[1]Here also, same phenomena happens in XRD analysis according to washing procedures and same explanation can be applied to that in the Table 1. However, with the same washing procedure W3 obtained samples except 4M200T12 series have fairly low crystallinity. The highest one is NT22, which shows peak height twice as high as one of P25. Selected sample, 4M200T12-(8/2w)SiO_2, has similar peak height to P25.

Table 4. Results from factor-level combinations of reaction conditions.

Factor & Level[1]			Step 1 (H_2 μmol, after 120 min)	Step 2 (H_2 μmol, after 120 min)
A	B	C		
0	0	0	0	0.28
0	1	2	0	1.052
1	0	1	4.107	0.949
2	0	2	5.493	0.69
0	2	1	0	0.652
1	1	0	0	3.981
1	2	2	2.17	14.628
2	1	1	11.403	7.814
2	2	0	0	6.885
Total			23.173	36.931

[1]Factor & Level combinations

Step 1: A=Pfu (Level 0= 0, 1=0.3, 2=0.6 Unit); B=Buffer (0=MES, 1=Tris, 2= Phosphate); C=Na-dithionite, 230 mM (0=0, 1=0.1, 2=0.2 ml)

Step 2: A=MV (Level 0= 0, 1=1.5, 2=2.5 mM); B=Tris (0=pH 7, 1=pH 8, 2= pH 9); C=Reaction Temp.(0=50, 1=60, 2=70 °C).

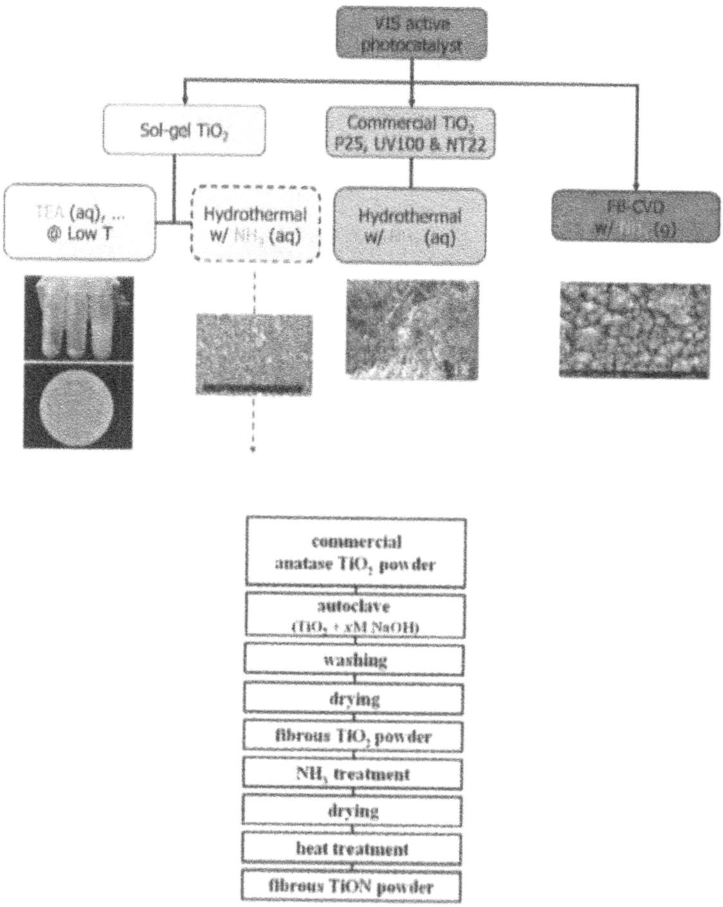

Figure 2. Flow chart for the preparation of various photocatalysts (top) and one of procedures of HT-TiON (bottom).

Figure 3. SEM images of optimized 4M200T12-8/2w after the NH_3 treatment (HT-TiON).

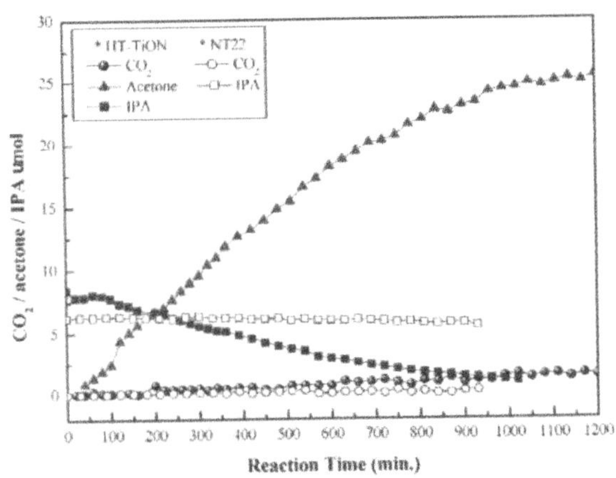

Figure 4. Photocatalytic IPA degradation using HT-TiON and NT22 (blue LED, 840 ml reactor, IPA=38 μmol, 0.15g of photocatalyst).

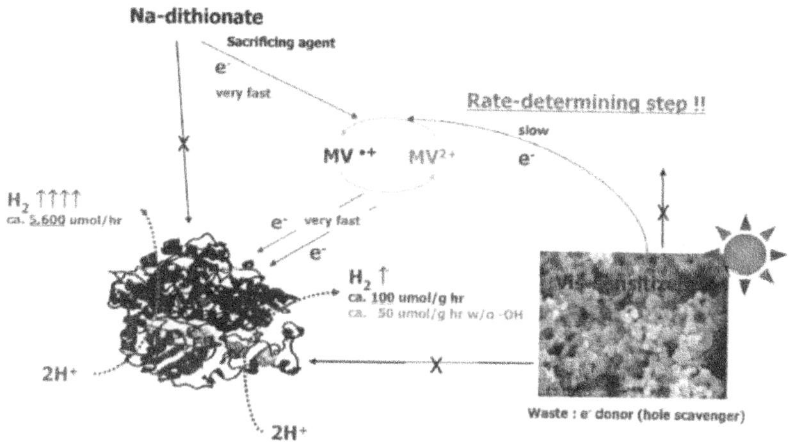

Figure 5. Schematic view of the mechanism for photo/biocatalytic hydrogen production system.

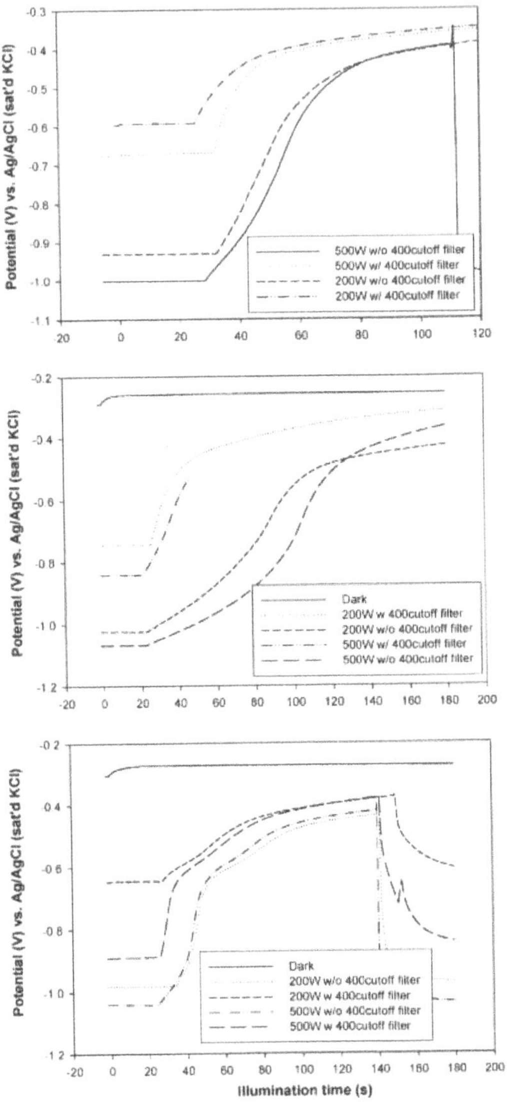

Figure 6. Potential vs. time curve using different light intensity and photoanodes with TiON450 (top), TiON400 (middle), and TiON300 (bottom).

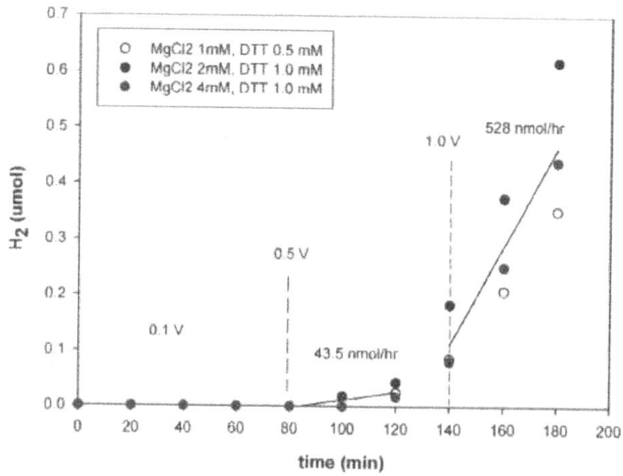

Figure 7. Change in the hydrogen production rate on external bias and injected chemicals.

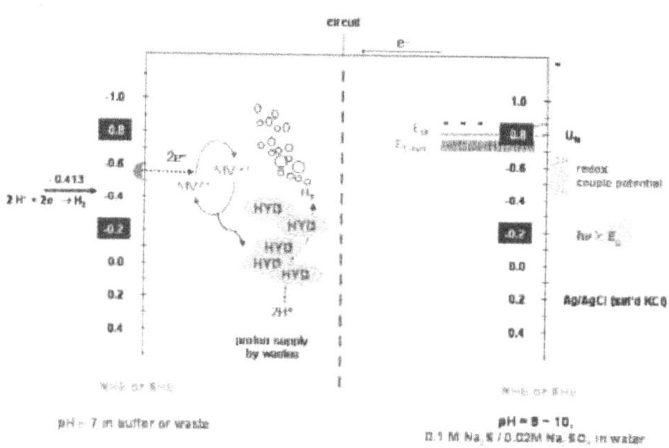

Figure 8. Expected reaction scheme for photo/biocatalytic hydrogen production system.

Eco-Materials Processing and Design
Edited by Koji Watari and Soo-Wohn Lee

EVALUATION OF TITANIA PHOTOCATALYST ADDED ZIRCONIA AND NITROGEN BY THE WATER PURIFICATION PERFORMANCE

Eiji Watanabe, Mitsuharu Fukaya, and Hiroshi Taoda
Ecomaterials Chemistry Research Group, Ceramic Research Institute,
National Institute of Advanced Industrial Science and Technology
(AIST) Chubu
Shimoshidami-Anagahora 2266-98, Moriyama-ku, Nagoya, 463-8560, Japan

ABSTRACT

We attempted to develop the new type titania photocatalyst that activated to respond in not only ultraviolet rays region but also visible light radiation region by the new method of the zirconia and nitrogen simultaneously introduced in the titania lattice. It was found that the amount of the nitrogen introduction was increasing by increasing of the amount of the zirconia addition in titania.

The degradation performance of new type titania photocatalyst fine particle was evaluated by water with the organic compound and compared with the conventional type under both the ultraviolet ray and visible light radiation conditions. It was found that the new type titania photocatalyst was also able to decompose the organic compound in water by the visible light radiation.

INTRODUCTION

The development of the environmental purification technology using the titania photocatalyst is remarkable, and the application is commercialized in various fields. However, the titania photocatalyst that is practical use mainly responds only to in the ultraviolet rays area, and the application range was limited. Also, the performance of the titania photocatalyst of the visible response type recently presented was not enough yet.

Then, we attempted to develop the new type titania photocatalyst that activated to respond in not only ultraviolet rays region but also visible light radiation region by the new method of the zirconia and nitrogen simultaneously introduced in the titania lattice for the application to the water purification treatment etc.

It is necessary to evaluate the degradation performance of new type titania photocatalyst fine particle by water with the organic compound for the application to the water treatment etc. However, the best test method to evaluate water quality purification ability has not been completed yet. We were already proposed the new evaluation test method of the water quality purification performance of the photocatalyst, that is, after degradation processing of the standard organic compound including the chlorine ion by the titania photocatalyst in water, the change of the chlorine ion concentrations that generated in the treating water was measured by the ion chromatography analysis method. We will report the degradation performance results of new type titania photocatalyst contracted with the conventional type by the above evaluation test method.

EXPERIMENT

Chemicals

10ppm solution of M-chlorobenzoic acid (ClC6H4COOH, m.w. 156.57) that is organic compounds of comparatively low molecular and contains chlorine ion uniting directly with the benzene ring of a basic frame is used as a standard.

Samples

To obtain the anatase type titania fine particle containing the zirconium dioxide (zirconia) for the range amount of 0.5 to 10 wt. %, the experimental procedure was operated as followings: Tetrachloride titanium solution was put into the 1000-ml tool container. The water-soluble zirconium salt solution (zirconium oxychloride octahydrate $ZrCl_2O \cdot 8H_2O$) containing the zirconia concentrations as compared with the above-mentioned weight ratio were gradually added to the titanium chloride solution at the range of the room temperature to 100°C. According to cases, it was necessary to do this procedure in the draft so that steam may diffuse along with the mixture. Then, it was neutralized the solution by adding the ammonium solution diluted by water (1:1) monitoring the methyl red indicator or pH meter. After standing sill over night, the precipitation of reaction product was separated by decantation or centrifuging (3,000rpm) to remove the supernatant liquid. It might be repeated the completely removing procedure of chloride ion at several times to add the same amount level of the purified water as precipitation at least. Then, dried at the room temperature, it was made to the fine particles by crushing gently. To temporarily burn the titania fine particles in the electric furnace, temperature was rising at a speed of 100°C per hour, and it maintained for two hours at 450°C. After cooled naturally, and the titania fine particles colored yellow was obtained. This process of manufacture is to make the zirconia and nitrogen addition titania photocatalyst by solid solution into titania crystal lattice during manufacturing the nitrogen substitutional titania as for zirconia. The nitrogen introduction is done by nitrogen's in ammonia uniting with titanium in the neutralizing reaction process to the titania with the aqueous ammonia. As the amount of the zirconia added to the titania, it is 0.5-10 percent by weight, and desirability is 1-5 percent by weight.

Contrast titania photocatalyst fine particles (ST21, produced by ISHHRA SANGYO KAISHA, LTD)

Apparatus.

Spectrum reflectivity of the titania fine particles were measured in the wavelength range of 380 nm to 780nm by the Hitachi Colorimeter (C-2000S).

XRD was used for identification of the crystal compositions and EDX for measurement of the chemical components in the producing materials.

Ion chromatography measurement systems and conditions were as follows: All chromatography was performed on a Dionex (Sunnyvale, CA, USA) 2010i type ion chromatography. Dionex manufactured all columns used in this study. Guard column: Ion Pac AG4-HC (50X2 mm). Analytical column: Ion PAC HPIC-ASA4-SC. IC Eluent: 2.25 mM Sodium carbonate and 2.8 mM Sodium bicarbonate. IC flow-rate: 1.0 ml/min. Detection: Suppressed conductivity, Anion Self-regenerating Suppressor ASRS II. Sample solution volume: 50 μ.

Analytical procedure

The constant amount (1g) of the titania photocatalyst fine particles was putting into the quartz glass square container (5cm X 5cm X 5cm), adding $10cm^3$ of the standard organic compound solution (10ppm concentration). The several samples were prepared. Their containers with quartz glass cover plate were placed left more than over one night in the shade box, and the adsorption equilibrium could be reached at room temperature (25°C).

One pair of the ultraviolet rays irradiation device (black light fluorescent lamp) and the fluorescent lamp that removed ultraviolet rays area were respectively installed in parallel. It was adjusted so that the ultraviolet ray or visible lights of constant strength (quantity of light of irradiation 1.0 mW/cm^2) might irradiate to the titania photocatalyst and the sample solution mixture in the containers from the upper side were lit in the shade box.

The container was taken out at arbitrary times during the irrigation procedure, and a treating solution in the container separated through the filter unit (pore size ϕ 0.2 μm. Millipore) was injected into the ion chromatograph. Chlorine ion generated from the organic compound degradation was identified and measured the concentration by the ion chromatograph analysis method.

RESULTS AND DISCUSSION

Chemical properties

Figure 1 shows that one of crystal composition of the new type titania photocatalyst fine particle that introduced zirconium and nitrogen was examined by the XRD. As for the peak that appeared in the XRD pattern, only the peaks based on the titania crystal of anatase type was identified and the zirconia crystal and titanium-zirconium complex crystallization etc. were not found at all in Figure 1. It is shown that the zirconia crystallization etc. was not generated to the titania crystal particle.

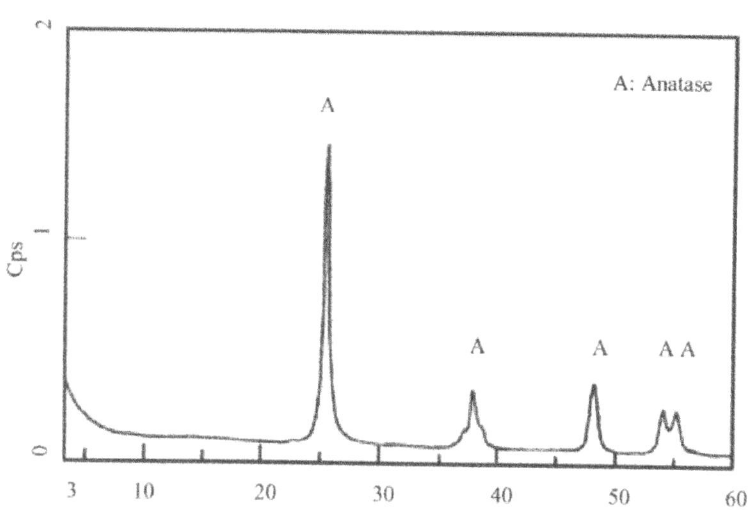

Figure 1. XRD pattern of the titania fine particles of nitrogen and zirconia introducing type (CuK∝)

It was thought that the zirconium atom added several % amounts and nitrogen were mostly solid-solved in titania crystal. It was thought that this depended on the difference of the amount of the nitrogen introduction in the titania that originated in the zirconium.

Chemical components of sample added only nitrogen and no zirconium and sample added to become arbitrarily weight % as zirconia for titania processed according with the developed synthesis method, were measured by EDX.

Figure 2 shows that the relationships between the content of the nitrogen dioxide and zirconia included in each samples were considered from respect of the weight concentration. As a result, it was appeared that the content of nitrogen dioxide increased as the content of zirconia increased in the titania fine particle. It was found that the amount of the introduction of nitrogen dioxide in the titania fine particle was able to be increased by adding simultaneously more than a constant amount of zirconia when titania was synthesized. However, the photocatalytic ability decreased when the content of zirconia exceeded 10% in the titania.•

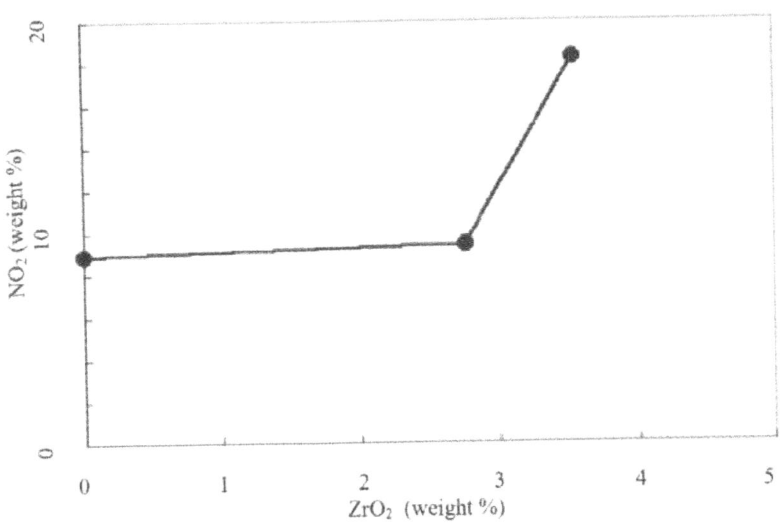

Figure 2. Relationships ZrO_2 contents and NO_2 induced into the titania

Table 1. Chemical composition of titania photocatalysts

Sample	Chemical composition
Original titania (anatase type)	TiO_2
Nitrogen doping type	$TiO_{1.85}N_{0.15}$
Zirconia and Nitrogen introducing type (new type)	$Ti_{0.9872}Zr_{0.0126}O_{1.69}N_{0.31}$

Table 1 shows chemical composition of synthesized samples. The chemical composition of a pure anatase type titania crystal was TiO_2. It was found that one of the types that doping nitrogen was $TiO_{1.85}N_{0.15}$. And, one example of new types of titania was $Ti_{0.9872}Zr_{0.0126}O_{1.69}N_{0.31}$.

The amount of nitrogen increased and the amount of oxygen decreased in one of the types that doping nitrogen relatively compared with the chemical composition of a pure anatase type titania crystal. It was found the nitrogen atom doping was replaced the oxygen atom in titania. The amount of titanium decreased in the new type to the extent that the amount of the zirconium increased. In addition, it was found that the amount of nitrogen increased to twice or more in the new type titania compared with the type of doping only nitrogen. It has been appeared that the amount of nitrogen replaced for oxygen in the titania crystal lattice increased by introducing the zirconium simultaneously.

The chemical composition of new type titania photocatalytic fine particle was different depending on a synthetic condition. As for the optimum range of the content of zirconia in the titania with a highly effective photocatalyst function, it was found that the range was from 3 to 5%.

Spectrum reflectivity of the titania fine particles

It was known well that the titania crystal fine particle used as a photocatalyst material didn't absorb light in the visible radiation region. This titania sample was colored to yellow thinly when was reducing by the gas (H_2 or CO) in the reactor at various temperatures (last report). The untreated titania fine particle (sample A in Figure 3) was able to absorb only the ultraviolet rays of 380nm or less. On the other hand, not only ultraviolet rays but also the visible range region of 400nm or more was admitted to absorb light as for the sample reduced (samples both D650 and D950). The visible radiation absorbed amount rose as following as the

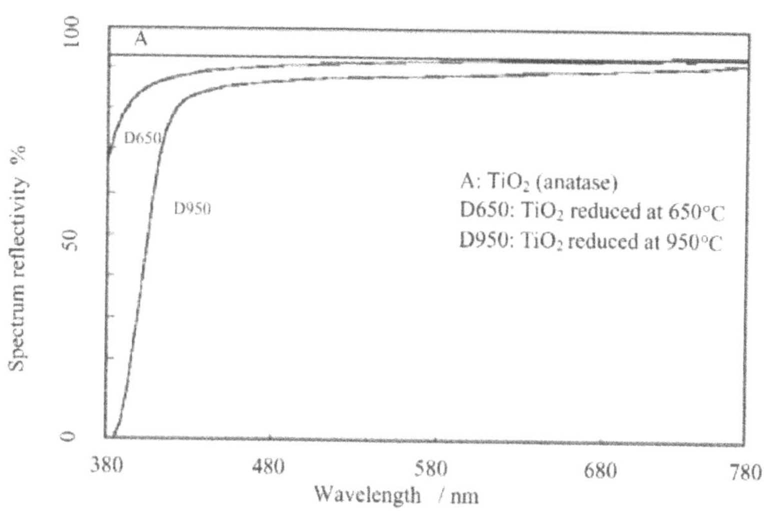

Figure 3. Spectrum reflectivity of the titania fine particles

temperature was increased. However, the range of the visible radiation absorption did not arrive to absorb the light of the wavelength of a narrow, more wide-ranging visible radiation region and there was a problem also in stability.

Spectrum reflectivity of titania synthesized.

To study the optical absorption property of the titania photocatalyst, the spectrum reflectivity of the obtained fine particles was examined. The photo absorbability can be numerical value converted according to the spectrum reflectivity.

Figure 4 shows the spectrum reflectivity of the titania fine particles synthesized for 2 hours heating at temperature 450°C. P marked the sample introducing nitrogen and no-zirconia titania, and Z marked the titania added 2wt. % as zirconia and nitrogen. The spectrum reflectivity of pure titania (anatase type) fine particle was only limited UV region, the wavelength 380nm or less. It was appear that both samples could absorb the light of the visible radiation region exceeding the ultraviolet rays region. It was found that Z sample absorbed more the visible radiation light in the range of 410nm to 450nm than P sample. Also, it was recognized that Z sample for the amount of the photo absorption was greatly large compared with the past value.

It was found that yellow coloring the titania fine particles became more vivid by adding nitrogen and the zirconium to the titania compared with the titania only introduced nitrogen. Then, to reveal the cause, the nitrogen content of both samples was examined by the heating experiment.

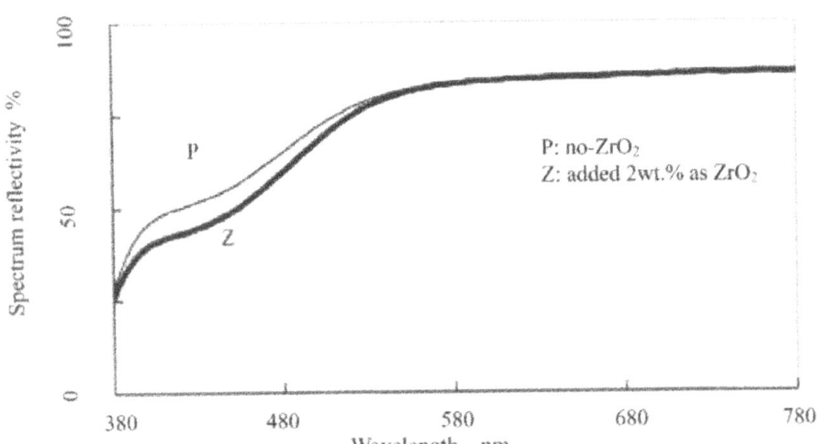

Figure 4. Spectrum reflectivity of the titania fine particles of nitrogen substitution type (treated for 2 hours heating at temp. 450°C)

After it had been completely oxidized heating both synthesized fine particles at 600°C, both samples were weighed respectively. It was found that the titania added the zirconium was larger for the weight change than the one not added it. It is appeared that the amount of the nitrogen introduced in the titania added the zirconium was more than the one not added it. It is thought that one of the causes that the amount of the photo absorption of titania added the zirconium increased compared with the one not added the zirconium is due to the difference of the amount of the nitrogen introduction.

The measurement results of the spectrum reflectivity of the titania fine particles that adds the zirconium and no-Zr one heat-treated for two hours at 450°C are shown. It is reported that the titania introduced only nitrogen absorbs the light of the visible range or more. Figure 4 shows that the absorption of Z-sample compared with P-sample in the range of 410nm to 450nm is great. In the newly developed Z-sample introduced simultaneously zirconia and the nitrogen, it was appeared that it could be expanding the visible region that was able to be absorbed further. A yellow taste of the sample has strengthened compared with the reduced type. The yellow degree of Z-sample is more vivid than others. It is shown that the developed material has a greatly large photocatalytic reaction in the visible radiation region compared with the past materials.

Evaluation of titania photocatalyst performance by the chlorine ion amount dissolved.

The chlorine ion released from the standard organic compound (chlorobenzoic acid) in the processing solution was analyzed with the ion chromatograph analysis. The chloride ion concentration in the solution that had been gathered every arbitrary time was plotted. Figure 5 shows the time series change of chlorine ion in the degradation process under the ultraviolet rays irradiation.

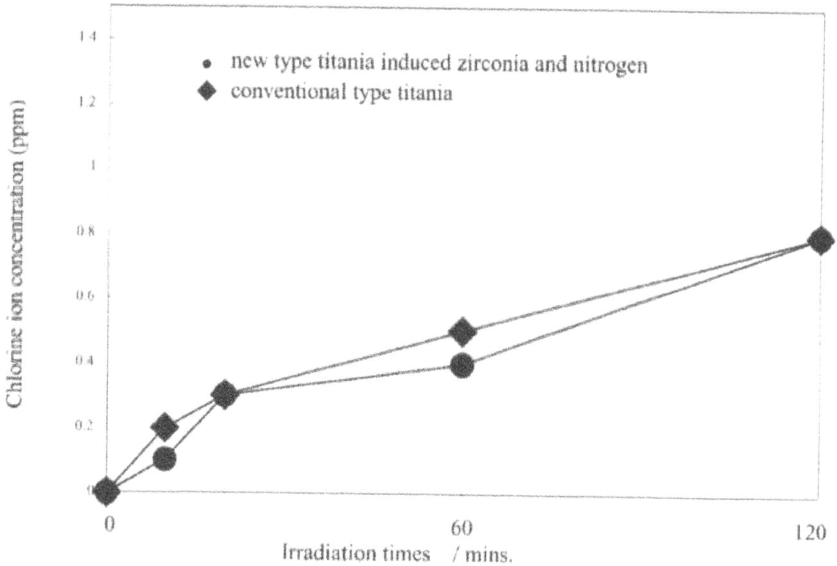

Figure 5. Change of chlorine ion concentrations from M-Chlorobenzoic acid resolved by titania photocatalyst under ultraviolet ray

The tendencies that the concentration of chlorine ion increases at time in the processing solutions were admitted. It is thought that the chlorine ion on a structural inside of chlorobenzoic acid is separated and released in the solution as decomposing by photocatalyst.

In case of the conventional type titania (as contraction) under UV irradiation, chlorine ion was increased in the processing solution with the passage of the irradiation time. In new type titania, it was also shown the chlorine ion generation tendency as well as in case of conventional type. In the both types, the organic compound was efficiently resolved by the photocatalyst action, and it is shown that the chlorine ion has separated in the solution as a result.

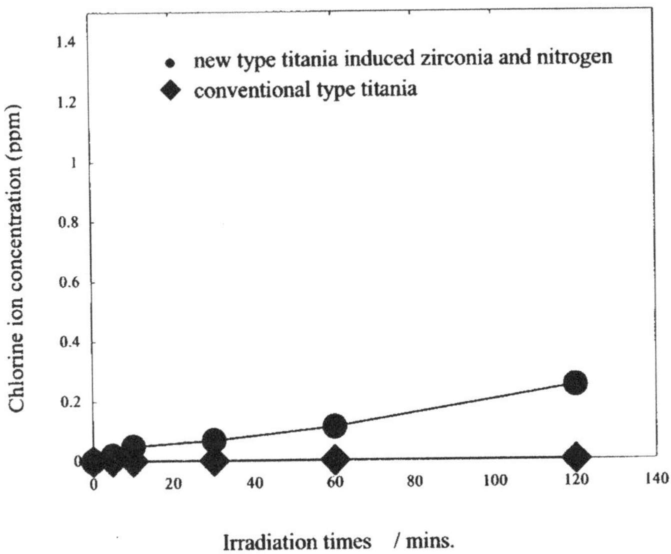

Figure 6. Change of chlorine ion concentrations from M-Chlorobenzoic acid resolved by titania photocatalyst under visible light

Figure 6 shows that the generation changes of chlorine ion were present in the solution processed under visible light irradiation. In case of conventional type titania, it was found that the chlorine ions in processing solution were not detected under visible light irradiation. It seemed that it has low photo catalytic activity under visible light irradiation. However, in case of the new type titania under visible light irradiation, chlorine ion was increased in the processing solution at time. It was found that the new type titania photocatalyst fine particle could resolve the organic compound in water under both UV and visible light irradiation. The ratio of the amount of chlorine ion released in the processing water to the it's total amount stoichiometically forecasted was about 5% at 60 minutes irradiation. As for the increase speed, however, a low tendency was admitted compared with the case under UV irradiation.

It was found that the generation extent of the chlorine ion was low somewhat value compared with the amount forecast by complete degradation of m-chlorobenzoic acid. It was found that it showed the possibility that a part of it has still changed into new organic compounds of a low molecule, and the chlorine ions was bonding with them because the ability of new type titania photocatalyst under visible light was low still compared with under UV irradiation.

CONCLUSIONS

The new type titania photocatalyst that introduced nitrogen and zirconium activated to respond in not only ultraviolet rays region but also visible radiation region. It was found that the titania fine particles colored to yellow, had also the photo-absorption ability in the visible range of 600nm to 400nm in addition to the range of ultraviolet rays. It was thought that using a photocatalytic reaction of new type titanium fine particle could become possible more widely by making to decomposition and harmlessness of organic compounds in water.

REFERENCES

[1] E. Watanabe, M. Fukaya, and H. Taoda, "Water Purification by Titanium Dioxide Thin Film Photocatalyst, " Reports *of NIRIN*, **46**, 227-236 (1997)

[2] E. Watanabe, M. Fukaya, K. Iseda, and H. Taoda, "Decomposition of organic substances in water by titanium dioxide photocatalyst, "*Proceeding of the 9th international conference on the Conservation and Management of Lake*, 169-172 (2001)

[3] E. Watanabe, M. Takagi, M. Fukaya, and H. Taoda, "Characteristic of titania photocatalyst fine particles added zirconia, "*Materials Sci. Forum*, **486-487**, 33-36 (2005)

[4] E. Watanabe, M. Fukaya, and H. Taoda, "Organism resolution ability of titania photocatalyst in water, "*Materials Sci. Forum*, **439**, 302-(2003)

[5] E. Watanabe, M. Fukaya, K. Tanaka, and H. Taoda, "Evaluation of titania photocatalyst revitalization by ion chromatograph analysis method, "*Proceeding of the 19th international Korea-Japan seminar on Ceramics*, 509-512 (2002)

[6] K. Iseda. A. Towata, E. Watanabe, H. Taoda, and M. Fukaya, "Relationship between the yields of products from ethanol over titanium dioxide and quantity of light," *Bull. Chem. Soc. Japan, 77, 1249-1253* (1998)

[7] E. Watanabe, T. Nonami, M. Fukaya, K, Iseda, and H. Taoda, "Changes of titanium oxide photocatalyst material by gaseous reduction, "*Preprints of the 1st Ceramics Environmental Materials Research Group Meeting*, 3 (2000)

[8] E. Watanabe, H. Taoda, T. Nonami, and M. Fukaya, "Adsorption and decomposition of bromine acid in water with reduced titanium oxide, "*Proceeding of the 4th international conference on ECOMATERIALS*, 527-530 (1999)

[9] M. Fukaya, T. Nonami, E. Watanabe, K, Iseda, M. Maeda, and H. Taoda, "Photocatalytic decolourization of dyeing waste water by TiO2 thin film coated on silicagel, "*Proceedings of the 3rd International Conference on Ecomaterials*, 28-30 (1997)

Eco-Materials Processing and Design
Edited by Koji Watari and Soo-Wohn Lee

PHOTOCATALYTIC PROPERTIES OF TIO_2 AND FE(III)-, Z(II)- AND SI(IV)- DOPED TIO_2 NANOPOWDERS SYNTHESIZED BY SOL-GEL

Qingfeng LI, Wei YUAN, Desheng Ai, Changsheng DENG and Xiaming DAI
Beijing Key Lab of Fine Ceramics and Institute of Nuclear and New Energy Technology, Tsinghua University
PO Box 1021, Beijing 102201, China

ABSTRACT

Nanosized TiO_2 and Fe-, Zn- and Si- doped TiO_2 powders were synthesized by sol-gel method and were characterized by X-ray diffraction (XRD), transmission electron microscopy (TEM) and diffuse reflectance spectroscopy (DRS). The photocatalytic activities of the synthesized nanopowders were investigated through the degradation of methyl orange solutions. The effect of the crystalline structure, crystal size and doping of different elements on the photocatalytic activities of the powders were discussedGood correlation between the diffuse reflectance spectra of TiO_2 powders and their photocatalytic activities were found: the smaller the particle size, the stronger the absorption in diffuse reflectance spectroscopy and the better the photocatalytic activities. Sol-gel synthesized TiO_2 nanopowders had higher photocatalytic activities than a commercial TiO_2 catalyst. Doping of TiO_2 with Fe, Zn and Si by sol-gel, however, led to reduced photocatalytic activities in comparison with a commercial TiO_2 powder. The reason for the reduction of the photocatalytic activities was discussed.

INTRODUCTION

It has become a focus of recent research how to improve the photocatalytic activities of TiO_2 [1]. The inherent photocatalytic efficiency lies with the catalytic sites available on the surface of the catalyst of a particular crystalline phase. By decreasing the particle size, hence increasing the surface area, the photocatalytic efficiencies can be improved. This can be achieved through methods such as sol-gel [2] or using inert particles, such as activated carbon [3], as template. The crystal structure of the titania catalyst plays a crucial role in determining the photocatalytic activities. For example, in the study of photocatalytic decomposition of naphthalene, Ohno *et al* [4] found that there exists synergistic effect between the two different crystalline forms of titania, i.e., anatase and rutile: presence of some small anatase particles is improving the photocatalytic activities of relatively larger rutile particles. This was explained by the transfer of electrons from rutile particles to anatase particles in the mixture system. Another way to improve the photocatalytic activity is by promoting the separation of electrons and holes in the photocatalytic processes. This can be achieved by doping, such as Ag^+ [5]. Their effects, although differ in different reaction systems, were mainly presumed to trap photo-induced electrons. However, there exists controversial effects regarding the doping to titania. According to Herrmann [6], doping would lead to detrimental effect to the photocatalytic activities. Other studies (see e.g., ref [7]) suggest that doping can improve the photocatalytic activities of titania. Therefore, there is still a need to investigate the factors affecting the photocatalytic activities so as to improve our understanding towards the photocatalytic properties of titania systems.

EXPERIMENTAL

0.05 mol $Ti(OC_4H_9)_4$ was dissolved in 1.5 mol anhydrous ethanol under stirring. Nitric acid was added to prepare a solution with its $[H^+]$=0.01 mol/L. To the solution 2 ml of deionised water was added dropwise and the mixture was continuously stirred for 30minutes. Then the mixture was left stand still to form a transparent orange gel. The gel was dried at 110°C to a white xerogel. Finally, nanopowders were obtained by heating the xerogel at 550°C for 2 h and 4h, respectively, labeled as TiO_2(2h) and TiO_2(4h) correspondingly. A commercial TiO_2 fine powder catalyst was also purchased, and labeled commercial TiO_2 fine powder herein.

Doping of 5%Fe(III) and 5%Zn(II) was realized by adding $Fe(NO_3)_3$ and $Zn(NO_3)_2$ respectively to $Ti(OC_4H_9)_4$ ethanol solution prior to hydration. Doping with 5%Si(IV) was carried out by hydrolyzing the mixed ethanol solution of $Si(OC_2H_5)_4$ and $Ti(OC_4H_9)_4$. The prepared gel was dried at 110°C for 24 hours and then heat treated at 600°C for various length of time to obtain the powders.

The crystalline structure of the powders was analysed by XRD, with a BRUKER D8 Advanced polycrystalline X-ray diffractometer with CuKα as the target. The particle morphology of the powders was observed using TEM with an H-800 Hitachi transmission electron microscope and the particle size was estimated with TEM micrographs of the powder particles. The specific surface area of the powders was measured with an ST-03 specific surface area analyzer. Diffuse reflectance spectroscopy was recorded using a UV2100 UV-Vis spectrophotometer.

The photocatalytic activities were studied using the degradation of methyl orange as a model reaction. Methyl orange, a typical azo-dye in the textile industry, [[(4-dimethylamino)phenyl]-azo]benzenesulfonic acid sodium salt with the molecular formula [$(CH_3)_2NC_6H_4N=NC_6H_4SO_3Na$]), is a very stable pollutant.

A 10 mg/L methyl orange solution (pH = 7) was first prepared, and 2 g of the commercial TiO_2 fine powders, 2 g of TiO_2(2h) and TiO_2(4h) each were then ultrasonically mixed respectively with the methyl orange solution under the illumination of a UV lamp (220V, 20W) and under constant stirring. After certain time interval, the specific UV-Vis absorption of methyl orange was measured and its degradation rate determined using the intensity of the band at 460 nm.

RESULTS AND DISCUSSION

XRD patterns of the powders (not shown) indicates that nanosized TiO_2(2h) is anatase and TiO_2(4h) is a mixture of anatase and rutile but mainly anatase. Based on the intensity of the peaks for anatase and rutile, the content of the rutile is estimated using the following formula [2]:

$$\%Rutile = \frac{1}{(0.844(A/R)+1)} \times 100\% \qquad (1)$$

where A and R are the peak area for the major anatase ($2\theta = 25.3°$) and rutile phase ($2\theta = 27.5°$), respectively. This gives a rutile content of 15 wt% in TiO_2(4h). The commercial TiO_2 fine powder consists entirely of anatase. There is line broadening in the XRD patterns of nanosized TiO_2 powders, more obvious with that of TiO_2(2h), which can be attributed to the fineness of the particles and a lower crystallinity of the nanosized powders than the commercial fine powder. The doped TiO_2 powders all consisted of anatase.

TEM micrographs (Fig 1) show that agglomerations exist in TiO_2(2h) powder but TiO_2(4h) powder particles appear to be better separated with fewer agglomerations. Elongation of the heating time from 2h to 4h apparently coarsened the particles. The specific surface area and the equivalent particle size of the three powders are summarized in Table 1. The particle sizes derived from specific surface area agree well with the TEM estimate. The commercial TiO_2 powder had the largest particle size. For TiO_2(2h) it is 8 - 10 nm; for TiO_2(4h) it is 40 – 50 nm; and the commercial TiO_2 fine powder has an average particle size of about 160nm. The TEM micrographs of the doped powders (Fig 2) also exhibited strong agglomeration and the sample doped with Fe had the strongest agglomeration, hence perhaps the largest secondary particle size.

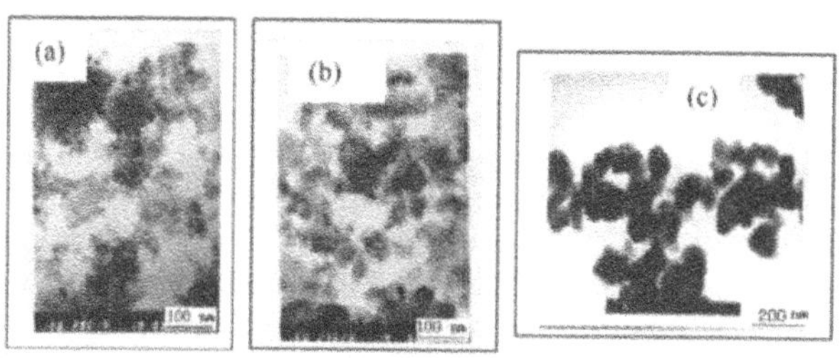

Fig. 1. TEM micrographs of TiO_2 powders obtained by heating the gel at 550°C for (a) 2h, and (b) 4h; (c) is the commercial TiO_2 powder.

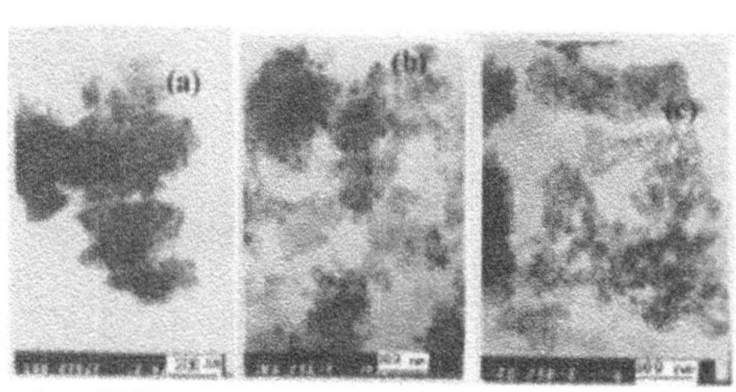

Fig. 2. TEM micrographs of TiO_2 nanopowders doped with (a) 5%Zn(II), (b) 5%Fe(III) and (c) 5%Si(IV).

Table 1 specific surface area and equivalent particle size of TiO_2 powders

Sample	Commercial TiO_2 powder	TiO_2(2h)	TiO_2(4h)
Specific surface area (m^2/g)	10.9	134.4	58.7
Equivalent particle size (nm)	200	16.4	37.1
TEM particle size (nm)	160	8 - 10	40 - 50

The band gaps for anatase and rutile are 3.2 eV and 3.0 eV, respectively, and their theoretical absorption edges, according to $\lambda < \lambda_g = 1240/E_g$, are at *ca.* 380nm and *ca.* 410 nm, respectively. Fig 3 indicates that the commercial TiO_2 powder has a diffuse reflectivity of 90% while in 360 - 380nm region it has a sharp absorption edge, suggesting an obvious semiconducting behaviour and corresponding to the anatase phase. TiO_2(2h) also has an absorption edge in the region 360-380nm, and in the region > 380nm its diffuse reflectivity is very small but increases slowly with the wavelength. This can be attributed to the nano-effect resulted from the very fine particle size. Still an absorption edge can be seen, suggesting its semiconductor characteristics and its crystalline structure is also anatase. TiO_2(4h) has an absorption edge in the region 365 - 410nm, which red-shifts in comparison with TiO_2(2h) and the commercial TiO_2 powders. Since the TiO_2(4h) powder has fine particle sizes, it also has some degree of nano-effect, and its diffuse reflectivity in the region > 410nm increases gradually with wavelength. This reveals different crystalline structure in TiO_2(4h). Two effects can be clearly observed from the diffuse reflectance spectra of TiO_2(2h) and TiO_2(4h): (1) the optical absorption of nanopowders increases rapidly with the decrease of the particle size; (2) the optical absorption beyond the absorption edge markedly decreases with the decrease of the photon energy.

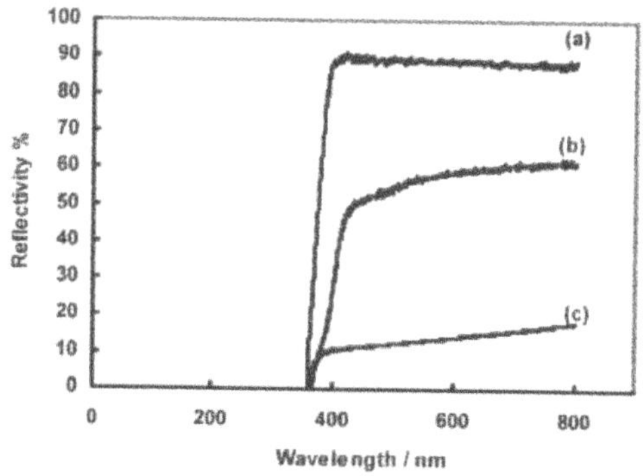

Fig. 3. Diffuse reflectance spectra for (a) commercial TiO_2 fine powder. (b) TiO_2(4h) and (c) TiO_2(2h)

The photocatalysis experiments were carried out in a neutral methyl orange solution, maintaining a constant pH before and after the reactions. According to Lambert-Beer's law, under a certain pH, the absorbance of methyl orange solution is directly proportional to the concentration of methyl orange in the solution. Fig 4 presents the plots of the concentration of methyl orange as a function of UV-Vis illumination time.

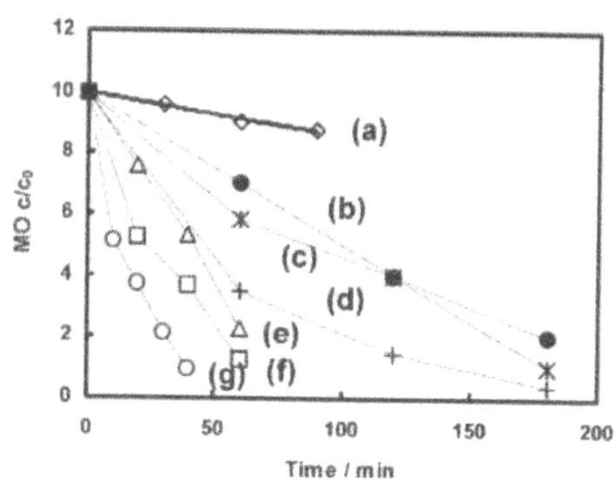

Fig.4. Concentration of methyl orange with time of UV illumination. (a) without TiO_2 added; and with TiO_2 added: (b) TiO_2 doped with Si(IV), Fe(III), (c)TiO_2 doped with Zn (II), (d) TiO_2 doped with Si(IV); (e) Commercial TiO_2. (f) TiO_2(4h) and (g) TiO_2(2h).

Fitting the photocatalytic reaction data (Fig 4) to the well-recognized first order kinetics associated with the adsorption and desorption process:

$$-\ln\left(\frac{C}{C_0}\right) = k't, \quad (2)$$

where C_0 and C are the methyl concentration at time $t = 0$ and t respectively during the degradation reaction and k' is the apparent first-order reaction constant, the reaction constants were derived and presented in Table 2. The kinetic data indicate that an order of decomposition rate exists: TiO_2(2h) > TiO_2(4h) > TiO_2 commercial > TiO_2-Si > TiO_2-Zn > TiO_2-Fe.

Table 2. Apparent reaction constant for the degradation of methyl orange without and with TiO_2 photocatalyst

Catalyst	None	TiO_2 commercial	TiO_2(4h)	TiO_2(2h)	TiO_2-Fe	TiO_2-Zn	TiO_2-Si
k' (s^{-1})	0.0015	0.0239	0.0329	0.0558	0.0084	0.011	0.0173

Without TiO_2 nanopowders, the degradation of methyl orange under UV radiation was negligible. The synthesized pure TiO_2 powders had higher catalytic activities than the commercial powder. Arabatzis *et al.* [8] recently studied the methyl orange degradation by gold modified TiO_2, and they obtained a decomposition constant for methyl orange catalyzed by TiO_2 film (Degussa TiO_2) of 0.0103 cm^{-1}. The present study gave a higher reaction constant, suggesting that the higher photocatalytic efficiency of the current reaction system. This may be attributed to the powder form of the present catalyst, whereas Arabatzis *et al*. used catalyst films, which could have fewer catalyst surface sites exposed to the reaction system.

That the TiO_2(2h) has the highest photocatalytic efficiency is consistent with the well-recognized opinion that anatase with fine particle size hence large specific surface area should exhibit high photocatalytic efficiency, particularly the fine size of the particles. If serious agglomeration exists among the nanoparticles, however, the electrolyte solution may not diffuse into the gap between the agglomerated particles, then the recombination probability of e^--h^+ will be greatly increased, and due to light shielding, their photocatalytic activities will be drastically reduced. Although TEM results indicated that the TiO_2(2h) has more agglomeration than TiO_2(4h), essentially the agglomeration led to the reduction of the surface area of the material, hence the reduction in available catalytic active sites and its effect on the photo-catalysis appears secondary.

The presence of rutile in the TiO_2(4h) powder clearly decrease the photocatalytic efficiency compared with the TiO_2(2h) powder which consists entirely of anatase. The effect of the increased particle size for TiO_2(4h), consequently the decreased surface area reduced the number of the active catalytic sites, leading to decreased photocatalyst efficiency. However, the higher photocatalytic activities of TiO_2(4h) than the commercial TiO_2 powders, which did not have rutile in its structure, suggests that the effect of the crystal structure, i.e., the reduced photocatalytic effect with the introduction of rutile might have been compensated for by the following two factors: (1) The smaller particle size of TiO_2(4h) than the commercial TiO_2 powder. Bacsa *et al.*[9] found that titania synthesized by sol-gel containing both phases shows a significantly higher catalytic activity than titania having only one of the two phases during the degradation of p-coumaric acid under light irradiation in the presence of H_2O_2. They suggested that the presence of the rutile phase introduced mesoporosity and a wider pore size distribution. These factors may be responsible for the increased catalytic activity. We do not have the pore size data to correlate this parameter with the catalytic activity. Nevertheless, it cannot be excluded that with the much smaller specific surface area (see Table 1) for TiO_2(4h) than for the commercial TiO_2 powder, the contribution of the pore size to the enhanced photocatalytic activities might also exist in the present system. (2) The conversion of the crystalline structure from anatase to rutile, which involves re-arrangement of the atoms, mainly oxygen, in the lattice, may induce the formation of some defects. The oxygen atoms may serve to arrest photo-induced electrons and could reduce e^{-}-h^{+} recombination probability.

The doped TiO_2 powders all exhibited some degree of photocatalytic activities, albeit smaller than the pure TiO_2 powders. The effect of doping is believed to be complicated. ZnO and Fe_2O_3 are semiconductors in their bulk. Doping with SiO_2 has been believed to retard the transformation of anatase to rutile. In the present work, the samples were so prepared that essentially the doped elements do not exhibit substantial crystalline structure at least to the level detectable by XRD. Therefore, it can be supposed that significant amount of the doped elements Zn and Fe exist in the powder as lattice substitution for Ti(IV) in TiO_2. This is supported by the XPS spectra for Ti2p3, as presented in Fig 5.

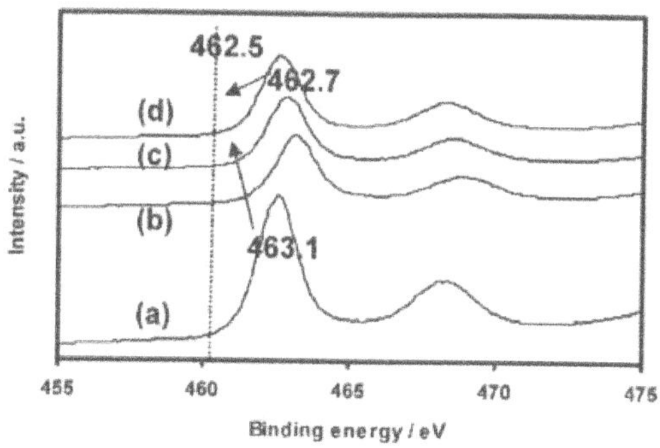

Fig. 5. XPS spectra for Ti 2p3 in (a) TiO_2, and TiO_2 doped with (b) Zn(II), (c) Fe(III) and (d) Si(IV).

The introduction of Zn and Fe led to the formation of defects in TiO_2:

$$ZnO \xrightarrow{TiO_2} Zn_{Ti}'' + V_O^{\bullet\bullet} + O_O \qquad (3)$$

$$Fe_2O_3 \xrightarrow{TiO_2} 2Fe_{Ti}' + V_O^{\bullet} + 3O_O \qquad (4)$$

The introduction of Zn_{Ti}'', $V_O^{\bullet\bullet}$ and Fe_{Ti}' and $V_O^{\bullet}$ solid solution defects caused the perturbation in charge distribution in the simple cubic TiO_4 cell: the Zn(II) or Fe(III) doped TiO_4 cell is negatively charged. This led to the charge in the neighboring cells to redistribute, and the decrease of the electron density of Ti2p3, hence the increase of binding energy of the core electrons. The substitution of Fe(III) into TiO_2 lattice has been proved possible with other technique [10]. It is noted that in the current study the Ti2p3 binding energy shift for Fe(III) dopedTiO_2 is less than that for Zn(II)-TiO_2. The difference in the shift in binding energy of Ti2p3 may partly be attributed to the valence difference between the doped elements and that of Ti(IV): an order of decrease exists - with that for Zn(II) the largest (2), followed by Fe(III) (1) and Si(IV) effectively has equal valence to that of Ti(IV). As the elements were doped into Ti04 and formed chemical bonds with Ti(IV), the greater the valence difference, the greater the perturbation caused to the outer shell electron density of Ti(IV) and so to its 2p3 electrons, and hence the greater shift in its binding energy.

As a result of the introduction of defects in the structure, these defects could be e^- - h^+ recombination centers. The introduction of Si into the TiO_2 structure also caused the decrease of the photocatalytic activities. The XRD for the Si doped TiO_2 suggests that the sample has lower crystallinity than the pure TiO_2. For Si(IV) doped TiO_2, the presence of crosslinking Si-O-Ti in the structure has been revealed [11] in sol-gel prepared SiO_2-TiO_2 samples, which could lead to the shielding of the surface active sites of TiO_2 catalyst, hence the reduced photocatalytic activities. Nevertheless, the Si- doped TiO_2 exhibited the highest photocatalytic activities among the three doped samples, perhaps the effect of the recombination centers provided by the Zn and Fe defects was more significant. Clearly further study on the doping level and physico-chemical characterization of the doped samples will reveal more information that could determine the photocatalytic performance of the doped materials.

CONCLUSIONS

The photocatalytic properties of TiO_2 are affected by the crystalline structure of TiO_2 and its particle size and doping of elements. Nanosization of TiO_2 anatase can improve the photocatalytic activities; Doping with Fe(III), Zn(II) and Si(IV) at 5wt% level via sol-gel route caused the formation of defects or Si-O-Ti crosslinking, leading to the decrease of the photocsatalytic activities of TiO_2. Further research is needed to acquire a better understanding of the relationship between the photocatalytic properties and the microstructure of the materials.

REFERENCES

[1] M. I. Litter, Heterogeneous photocatalysis transition metal ions in photocatalytic systems, *Applied Catalysis B: Environmental* **23**, 89 (1999).
[2] K. J. Jun and S. B. Park, Anatase-phase titania: preparation by embedding silica and photocatalytic activity for the decomposition of trichloroethylene, *J. Photochem. Photobiol. A: Chemistry*, **127**, 117(1999).
[3] G. Colón., M.C. Hidalgo, J.A. Navio, A novel preparation of high surface area TiO_2 nanoparticles from alkoxide precursor and using active carbon as additive, *Catalysis Today* **76**, 91-97 (2002).
[4] T. Ohno, K. Tokieda, S. Higashida, M. Matsumura, Synergism between rutile and anatase TiO_2 particles in photocatalytic oxidation of naphthalene, *Appl. Catalysis A: General*, **244**, 383(2003).
[5] H. Tsuji, H Sugahara, Y. Gotoh, J. Ishikawa, Improvement of photocatalytic efficiency of rutile titania by silver negative-ion implantation, *Nucl. Instrum. Methods Phy. Res. B* **206**, 249(2003).
[6] J. M. Herrmann, Heterogeneous photocatalysis: fundamentals and applications to the removal of various types of aqueous pollutants, *Catalysis Today* **53**, 115(1999).
[7] Z. H. Yuan, J. H. Ji, L. D. Zhang, Influence of co-doping of Zn(II) + Fe(III) on the photocatalytic activity of TiO_2 for phenol degradation, *Mater. Chem. Phy.* **73,** 323 (2002).
[8] I. M. Arabatzis, T. Stergiopoulos, D. Andreeva, S. Kitova, S.G. Neophytides, and P. Falaras, Characterization and photocatalytic activity of Au/TiO_2 thin films for azo-dye degradation, *J. Catalysis* **220**, 127(2003).
[9] R. R. Bacsa, J. Kiwi, Effect of rutile phase on the photocatalytic properties of nanocrystalline titania during the degradation of p-coumaric acid, *Appl. Catalysis B: Environ.* **16,** 19(1998).
[10] Z.M. Wang, G. X. Yang, P. Biswas, W. Bresser, P. Boolchand, Processing of iron-doped titania powders in flame aerosol reactors, Powder Technology, **114,** 197(2001).
[11] G. M. Ingo, C. Riccucci, G. Bultrini, S. Dire, G. Chiozzini, Thermal and microchemical characterisation of sol-gel SiO_2, TiO_2 and $SiO_{2-(1-x)}TiO_2$ ceramic materials, *J. Therm. Anal. Calorim.*, **66**, 37-46(2001).

Eco-Materials Processing and Design
Edited by Koji Watari and Soo-Wohn Lee

STUDY ON TITANIUM OXYNITRIDES BY HYDROTHERMAL METHOD USING A FRACTIONAL FACTORIAL DESIGN AND FLUIDIZED BED CHEMICAL VAPOR DEPOSITION (FB-CVD)

Hyunku Joo, Jaehyeon Park, Seung Yong Lee
Korea Institute of Energy Research (KIER)
71-2 Jang-dong, Yusong-ku
Daejeon, 305-343 Korea

Jinwook Ha
Soonchunhyang Univeristy
#646 Eupnae-Ri, Shinchang-Myun
Asan-Si, 336-745 Chungnam, Korea

ABSTRACT

This study examined the possible role of N-substituted TiO_2 (HT-TiON), which was prepared using a hydrothermal method with a fractional factorial design at the initial stage in order to classify the reaction conditions, when visible light or solar light is absorbed to initiate photocatalysis. The HT-TiON was obtained by treating a TiO_2 (HT- TiO_2) sample with a BET surface area over 100 m^2/g, prepared from a commercial TiO_2 powder (NT22), with aqueous NH_3 followed by heat treatment. The photocatalytic activity in IPA degradation under blue LED (peaked at 464 nm) irradiation was performed and compared with another sample; NT22. HT-TiON produced acetone. NT22 did not produce it even though it showed higher rate than HT-TiO_2 under UV irradiation.

In this work, TiON nano-coated photocatalysts on beads were prepared by FB-CVD method. Prepared photocatalysts on beads were characterized by FE-SEM/EDAX, and the photocatalytic activity in gas phase was studied with IPA under blue LED using GC. The particles deposited by nitrogen doped titania photocatalyst showed visible-light induction. It was confirmed in gas phase by the decomposition of IPA to CO_2 and H_2O.

INTRODUCTION

Environmental remediation using photocatalysts has been a major focus in many countries over the last decade. However, the photocatalytic process has been simultaneously criticized for being uneconomical compared to other oxidative treatment systems due to its limitations resulting from the necessities of an appropriate light source and immobilization, which may increase the overall energy costs[1]. Therefore, there has been a great deal of interest in tackling the aforementioned key issues to economically commercialize photocatalytic systems. One of the necessary conditions is activity in visible light (solar light, fluorescent lamp, etc.) as an activation energy source in photocatalysis. For the photocatalyst particles to be activated by visible light, nitrogen-substituted TiO_2 (TiON) has been studied, either by sputtering or by hydrolyzing aqueous $Ti(SO_4)^2$ with a NH_3 solution[2-3]. In addition, the difference in activity is not only related to their absorbance spectra, but the difference also depends on the physical properties of the samples such as the shape, distribution pores, crystalline phase, crystallinity, and the preparation and post-deposition treatments. Hence, in order to prepare visible light sensitized TiON, a hydrothermal technique has been considered as an experimental tool, which is known as fractional factorial design. The hydrothermal technique has been most popular, particularly in the last fifteen years, and has attracted a great deal of interest from scientists and technologists. The term hydrothermal is simply used to describe the action of water (or aqueous solvents) at elevated temperatures and pressures to bring changes in materials. The selected TiON sample in any shape might be combined with a biocatalyst (or hydrogenase enzyme) in the near future to produce hydrogen, which is currently a major research focus in energy and environment-related research.

FB-CVD is a useful method in catalyst preparation for the improvement of physicochemical properties such as catalytic activity, hardness, and mechanical stability since it deposits nano sized particles of the catalytic compound possibly on selective sites of the substrate [4-5]. This process can uniformly coat metal or metal oxide on small, three-dimensional objects such as beads, powders, fiber, and small pieces of equipments, which is not possible in conventional chemical vapor deposition, because entire surface area is not exposed to the activated gas.

The objectives of the work were to develop the FB-CVD (fluidized bed chemical vapor deposition) technology and to prepare the visible-light induced photocatalyst nano-coated particles using FB-CVD. FB-CVD is a new conceptual technology relating NT (nano-technology) and ET (environmental technology). The thickness of the film can be easily controlled. The photocatalyst is expected to be widely used in environment and energy fields with high efficiency, visible-light induction, and stable fixations. Therefore it is important to develop FB-CVD technique and produce photocatalyst nano-coated 3-D particles having visible-light induction.

EXPERIMENTAL

Hydrothermal Method

Visible light-activated nitrogen-substituted titanium oxide (TiON) was prepared using the hydrothermal method. P25 TiO_2 (Degussa) and NT22 TiO_2 (NANO Co., Ltd. Korea) were also used as a reference when necessary. In a typical preparation procedure for TiON, anatase TiO_2 powders were placed into a teflon-lined autoclave with a 50m capacity. The autoclave was then filled with an aqueous solution of NaOH with a molar concentration up to 80% of the total volume of the autoclave and was maintained under the desirable conditions without shaking or stirring during the heating process. Washing and drying was then performed. In this case, the one-third fractional factorial experiment involving 9 FLC was used to reduce the size of the experiment: a full factorial experiment a 3^3=27 factor-level combination (FLC) occurs when there are three factors and three levels for each factor. The selected factors consist of the NaOH concentration (levels of 1, 3 and 5M), temperature (levels of 175, 220 and 225 □) and time (levels of 12, 24 and 36 hr). As a preliminary experiment, raw TiO_2 for TiON was prepared using titanium tetraisopropoxide (TTIP), isopropyl alcohol (IPA), acetic acid and water for comparison. Finally, the nitrogen into the hydrothermally treated TiO_2 was obtained by mixing the dry TiO_2 powder with a NH_3 solution followed by drying (Fig. 1). The light sources used were UV-A lamps (Sankyo Denki, Japan) and LED (green and blue peaked at 520 nm and 464 nm, respectively.). Isopropyl alcohol (IPA) was a probe compound for measuring the activity. The chemicals (IPA, acetone and CO_2) were analyzed by gas chromatography with a flame ionization detector (GC/FID, HP 5890) and a GC/FID with a methanizer (HP6890). The structure, morphology and the surface area were investigated using XRD (Rigaku, DMAX/2000-Ultima Plus), SEM (Topcon SM-720), and BET.

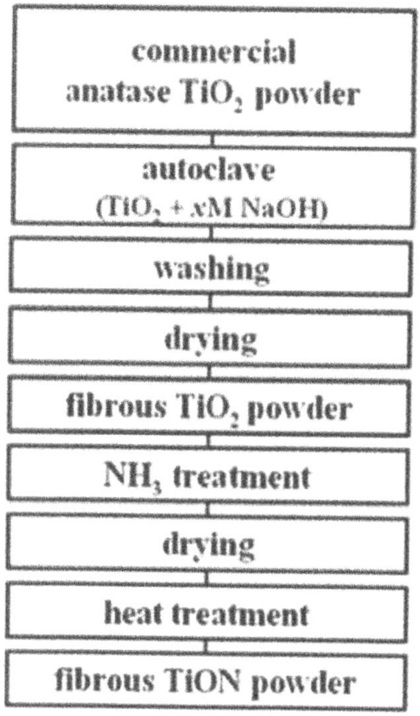

Fig. 1. Experimental procedure for HT-TiON.

Fluidized Bed Chemical Vapor Deposition

TiON photocatalysts were prepared by FBCVD method and titanium tetra iso-propoxide (TTIP, $Ti[OCH(CH_3)_2]_4$, Aldrich Chemical Co.) was used as a precursor of titania. Also, alumina beads were used as substrates. Fig. 2 shows the Fluidized Bed Chemical Vapor Deposition (FB-CFD) apparatus (SUS tube: 40 mm i.d. and 0.5 m height) used for the deposition of N-substituted titanium dioxide photocatalyst on substrate. The fluidizing gas is supplied to the bottom of the riser. (In circulating fluidized bed, the particles leave the riser through an horizontal tube at the top, and are separated from the gas in a cyclone. They are recirculated to the riser through downcomer and L-valve. The flow rate of the aeration gas to the L-valve is the main parameter to control the solid flux. The gas exit of the cyclone is linked to the vacuum unit.) Mixtures of argon, oxygen, NH_3 and TTIP were fed as reaction gases at vacuum and high temperature conditions. Temperature and pressure were measured above and bottom of the riser.

In the thermal CVD of thin TiON films, TTIP is decomposed in the fluidized bed and react with oxygen and NH_3 on the surfaces of substrate to form thin films of TiON. The films are deposited on the beads and reactor wall. A total mass of 300 g bead particles was handled per batch.

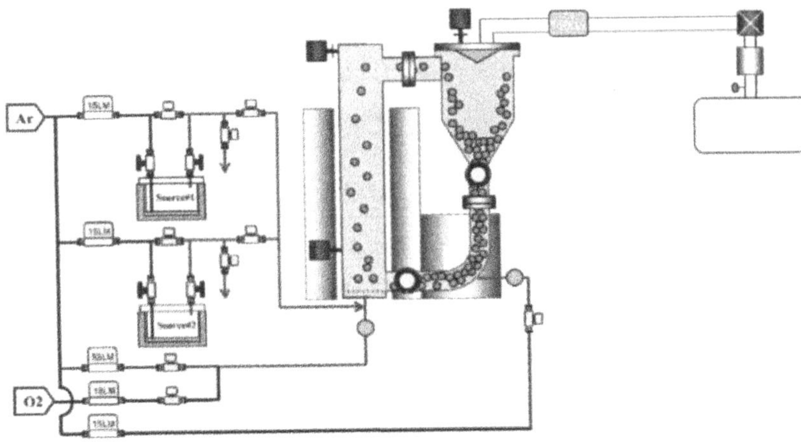

Fig. 2. Schematic of Fluidized Bed Chemical Vapor Deposition (FB-CVD).

RESULTS AND DISCUSSIONS

Hydrothermal Method

Initially, a rough hydrothermal experiment with NH4OH and the prepared raw TiO_2 (189 m^2/g) revealed that the sample turned into a yellow color and absorbed light with wavelength < 530 nm. In addition, using aqueous NaOH (10 M) and a 12hr reaction time under varying reaction temperatures ranging from 100 ~ 200°C 5 samples (with 25°C interval) exhibited the beta phase TiO_2 (β-TiO_2; Na_2TiO_3, $Na_2Ti_3O_7$) from XRD analysis, showing a maximum specific surface area of 246 m^2/g at 175°C. The addition of silica made the particle size more uniformly distributed (a fibrous shape was identified by SEM, Fig. 3). The tube shape TiO_2 using the hydrothermal method was identified and explained by Wang et al.[5], where they suggested a mechanism by which anatase TiO_2 nanotubes grow. They also reported that the nanotubes were formed during the alkali treatment, with subsequent acidic treatments having little or no effect on nanotube structure and shape. In order to investigate the effect of the washing procedure on the physical properties of selected samples (HT- TiO_2 from commercialized TiO_2 powder (NT22) under 10M NaOH, 175°C, 12hr), four different washing procedures were performed (W0: without washing, W1: washing only with water, W2:washing only with HCl, W3: washing with water and HCl). The following properties: specific surface area, pore radius and pore volume, were changed with respect to the washing procedure used, showing a linear relationship between the specific surface area and pore radius (Table 1). XRD showed that NaOH remained on the surface without washing (W0), with the other three methods showing no difference (β-TiO_2 present). This suggests that washing is essential for removing impurities and controlling the shape.

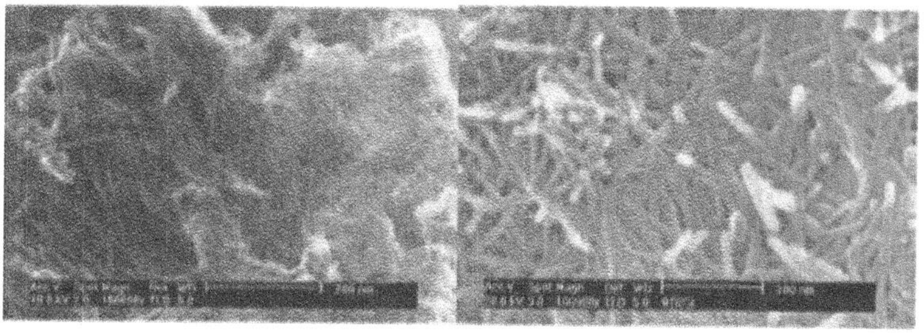

Fig. 3. SEM images of the hydrothermal prepared TiO_2.
(top: 10M NaOH, 175°C, 12 hr, bottom: with 20 vol% SiO_2 added)

Table 1. Effect of the washing procedure[1] on the physical properties of HT-TiO_2 from NT22 under 10M NaOH, 175 °C, 12hr

	BET surface area (m^2/g)	Pore volume (cm^3/g)	Pore radius (nm)
NT-22	64.7427	0.22792	6.82975
W0	16.0104	0.05916	7.38978
W1	164.3932	0.33846	4.11769
W2	308.8273	0.85506	5.53749
W3	308.8627	0.88443	5.72702

[1]W0: without washing, W1: washing only with water, W2: washing only with HCl, W3: washing with water and HCl

A systematic approach for preparing the photo-active TiO_2 was made using a fractional factorial design (FFD) with the following results. Table 2 shows the selected 9 factor-level combinations (FLCs) out of the total of 27 FLCs. From an analysis of the variance (ANOVA), reaction temperature had 99% level of significance on the specific surface area. The NaOH concentration was critical to the phase formed. For example, the anatase phase was formed with 1M NaOH, the β phase was formed with 5M NaOH, and the anatase was formed with below 225°C and β phase above 225°C for 3M NaOH. The specific surface area increased from 65 m^2/g to 144 m^2/g (5M NaOH, 225°C, 12hr). Fig.4 shows SEM images for selected samples, where a fibrous shape of TiO_2 was revealed at concentrations > 3M of NaOH. Detailed investigations based on the 3M NaOH, 175°C and 12hr were examined in terms of the phase and shape, varying the NaOH concentration (3.5, 4, 4.5 M) first. This was followed by variation of the reaction temperature and time. The resulting optimum condition was found to be 4M of NaOH, 200°C and 12 hr (103 m^2/g). A summary of the physical properties of the selected samples is shown in

Table 3, and their photocatalytic activity in degrading IPA using UV-A irradiation is shown in Fig. 5. P25 showed the fastest reaction rate, followed by NT22. The optimized sample (4M200T12-8/2w) showed a better activity than the other samples (4M200T12-), but was less active than the two commercialized TiO_2 samples under UV-A irradiation.

Table 2. Selected factor-level combinations (FLCs) and ANOVA analysis for the significant factor selection on surface area of TiON

	1M NaOH			3M NaOH			5M NaOH		
	12hr	24hr	36hr	12hr	24hr	36hr	12hr	24hr	36hr
175 °C	•				•				•
200 °C			•	•				•	
225 °C		•				•	•		

Source[a] (df)	SS	MS	F_o
Total (8)			
A (2)	93.639	46.819*	
B (2)	1016.992	508.496	11.71[b]
C (2)	34.657	17.329*	
A×B (2)	132.156	66.078*	
Residual		43.409 (6df)	

[a] : A: mole of NHOH, B: Rxn temperature, C: Rxn time, * : Components involved in residuals
[b] : 0.01 level of significance.

Table 3. Physical properties of the selected samples

	P25	NT22	4M200T12-(8/2w)SiO_2	4M200T12	3M225T36	5M225T12
Phase	anatase+rutile	anatase	anatase	anatase	anatase+ β-TiO_2	β-TiO_2
XRD intensity (θ)	545 (25.18)	1392 (25.16)	688 (25.30)	528 (25.26)	454 (25.26)	262 (24.26)
1/2 width	0.36	0.42	0.43	0.43	0.46	-
BET (m^2/g)	50	64.7	127.6	102.7	52.3	114.0
Pore Vol. (cm^3/g)	-	0.228	0.448	0.365	0.189	0.566

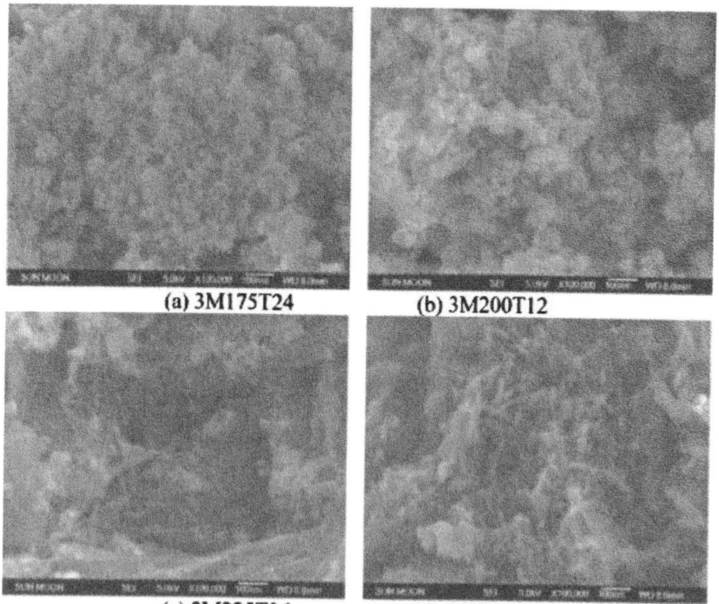

(a) 3M175T24 (b) 3M200T12

(c) 3M225T36 (d) 5M175T36

Fig. 4. SEM images of selected samples from the FFD experimental design.

The optimized sample (4M200T12-8/2w) was further analyzed by DSC to identify the heat-treatment temperature, where a distinguishable exothermic band appeared near 430°C (Fig. 6). This exothermic reaction was attributed to be the oxidation reaction of NH_3 or NH2 with the oxygen released from the amorphous grain boundaries by forming oxygen-deficient sites. This band was not significant due to the smaller amount of the amorphous portion in the obtained samples by the hydrothermal method. As a consequence, NH_3 treatment, followed by heat-treatment at 450°C produced yellow powders (HT-TiON, Fig. 7). A commercially available blue LED (peaked at 464 nm) was used to check the photocatalytic activity of the obtained HT-TiON (Fig. 8). Acetone evolution as a function of blue LED irradiation time was remarkable for the HT-TiON sample, but no acetone was detected for the NT22 sample.

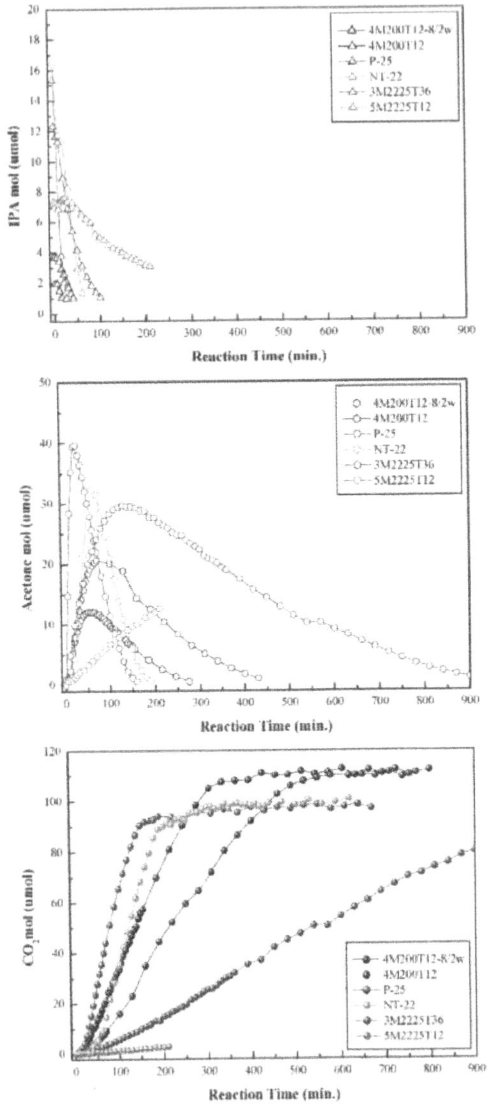

Fig. 5. Photocatalytic IPA degradation using selected samples
(UV-A lamp, 840 ml reactor, IPA=38 μmol, 0.15g of photocatalyst).

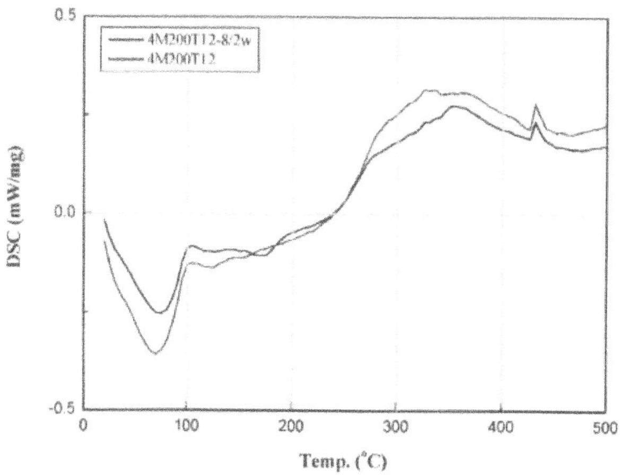

Fig. 6. DSC curve for 4M200T12 & 4M200T12-8/2w as a function of the calcination temperature.

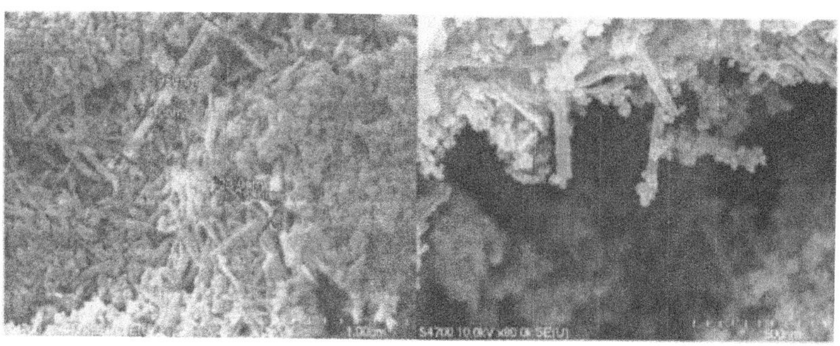

Fig. 7. SEM images of 4M200T12-8/2w after the HN_3 treatment (HT-TiON).

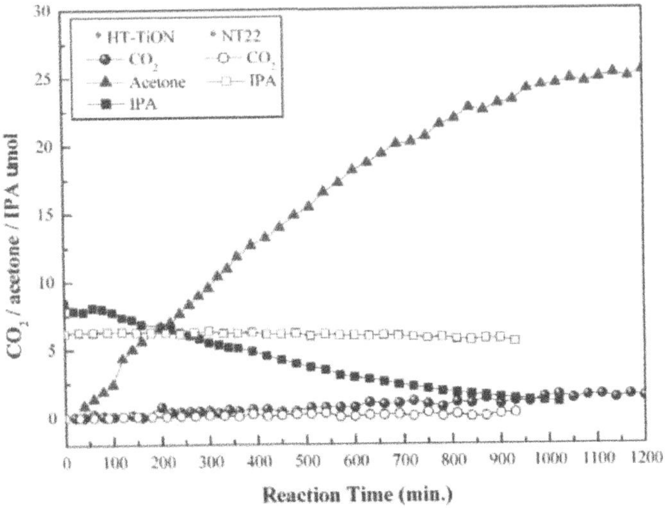

Fig. 8. Photocatalytic IPA degradation using HT-TiON and NT22 (Blue LED, 840 ml reactor, IPA=38 μmol, 0.15g of photocatalyst).

Fluidized Bed Chemical Vapor Deposition

The FE-SEM images of the surfaces of titania nano-coated alumina beads, silica-gel beads, glass beads and cross section of glass beads prepared by FB-CVD are shown in Fig. 9. From the FE-SEM photographs, it was found that the surfaces of titania nano-coated on beads were covered with crystal form, particle form, and slick form of titania on alumina, silica-gel, and glass beads, respectively. From (d) of Fig. 9, it could be verified that the thickness of coated titania was about 35nm and titania was evenly distributed on the surface.

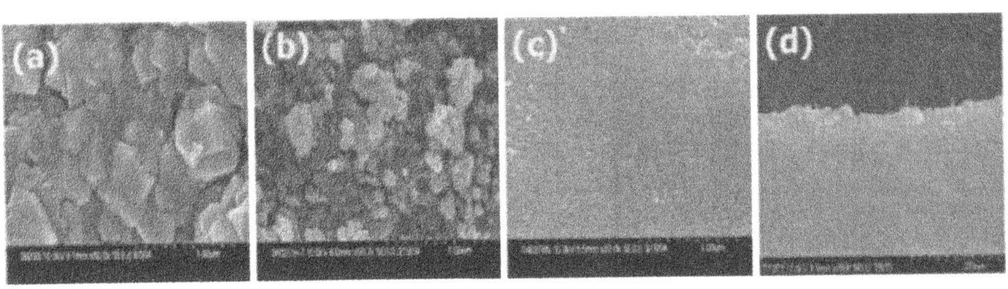

Fig. 9. FE-SEM images of surfaces of (a) TiO_2 /alumina, (b) TiO_2 /silica-gel, (c) TiO_2 /glass and cross section of (d) TiO_2 /glass.

In Fig. 10, The photocatalytic activities of TiO_2, Pt TiO_2, TiON(NH_3: 25, 50, 75cc/min) on beads under Blue LED are shown. It is seen that the reactivity of TiON (50 cc/min of NH_3) is higher than others and Pt TiO_2 on the bead shows lowest activity.

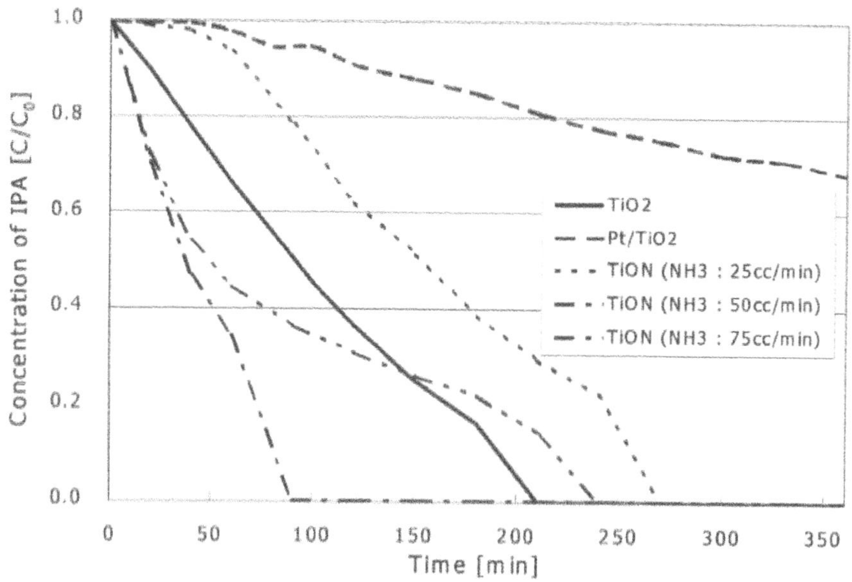

Fig. 10. Photocatalytic reactivities of TiO_2, Pt TiO_2, TiON(NH_3: 25, 50, 75cc/min) on alumina under Blue LED.

CONCLUSION

This study investigated the role of N-substituted TiO_2 to initiate photocatalysis with visible light. With a limited loss in the reduction potential, N-substituted TiO_2 can play a major role in visible light sensitized photocatalysis in both fields: environment purification and hydrogen production with hydrogenase, which is a major research focus in this group. A highly porous visible light sensitized material is essential for efficiently providing the hydrogenase excited electrons needed to produce hydrogen without the need for an electron mediator or donor.

The visible-light induced photocatalyst nano-coated particles were successfully prepared with FB-CVD technique. The particles deposited by nitrogen doped titania photocatalyst showed visible-light induction. It was confirmed in gas phase by the decomposition of IPA to CO_2 and H_2O under BLED.

ACKNOWLEDGEMENTS

This research was performed for the Hydrogen Energy R&D Center, one of the 21st Century Frontier R&D Program, and partly for KIER R&D program, both funded by the Ministry of Science and Technology of Korea, and Ministry of Environment of Korea.

REFERENCES

[1]Raupp GB, Alexiadis A, Hossain M, Changrani R., Catalysis Today, 69 (2001) 41-42.
[2]Ashai R, Morikawa T, Ohwaki T, Aoki K, Taga Y., SCIENCE, 293 (2001) 269-271.
[3]Ihara T, Miyoshi M, Iriyama Y, Matsumoto O, Sugihara S., Applied Catalysis B: Environmental, 42 (2003) 403-409.
[4]S. Morooka, T. Okubo and K. Kusakabe: Power Tech., 63(1990), 105-112.
[5]S. J Lee, M. H. Kim, Y. T. Kim and G. Y. Chung: Korea J. Chem. Eng., 19 (2002), 167.
[6]Wang, W., Varghese, O. K., Paulose, M., Grimes, C. A., Wang, Q., Dickey, E. C., J. Mater. Res., 19(2) (2004) 420-421.

Eco-Materials Processing and Design
Edited by Koji Watari and Soo-Wohn Lee

CORRELATION BETWEEN ANATASE-TO-RUTILE PHASE RATIO TO PHOTOCATALYTIC ACTIVITY OF TiO_2 OBTAINED BY SOL-GEL METHOD

Kannikar Juengsuwattananon and Supatra Jinawath
Department of Materials Science, Faculty of Science, Chulalongkorn University, Phayathai Rd, Pathumwan, Bangkok, 10330 Thailand

Sitthisuntorn Supothina*, Angkhana Jaroenworaluck and Thammarat Panyathammaporn
National Metal and Materials Technology Center, 114 Thailand Science Park, Paholyothin Rd., Klong Luang, Pathumthani, 12120 Thailand. *E-mail:sitthis@mtec.or.th

ABSTRACT

The effect of phase composition of the TiO_2 particles was investigated. Various ratios of anatase and rutile were obtained by controlling the synthesis parameters such as the ratio of starting reactants, *i.e.* tetrabutyl orthotitanate, ethanol, water and nitric/hydrochloric acid, as well as aging times. Heat treatment of the pure anatase phase was also conducted to vary the phase ratio. Correlation between phase ratio and photocatalytic activity was studied by measuring methylene blue decomposition under UV illumination. The activity result was compared with that of the commercial Degussa P-25. The effect of synthesis parameter on phase formation of the TiO_2 has also been described.

INTRODUCTION

Photocatalytic reactions by metal oxide semiconductors have been widely investigated in the degradation of toxic organic/inorganic pollutants, energy conversion and storage.[1-4] Among various oxide semiconductor photocatalysts, TiO_2 is one of the most popular and promising materials in photocatalytic application due to its high activity, long-term stability to light illumination, low price and nontoxicity. Typically, the TiO_2 is present in three natural forms; anatase, rutile and brookite. Phase structure, crystallite size, morphology and specific surface area of the TiO_2 have been reported to relate to photocatalytic efficiency.[5-8] The anatase has been reported to be the most active form although the mixed phase of the anatase and rutile was also found to be active as in the case of a commercial Degussa P-25. Typically, the ratio of anatase and rutile is varied by controlling heat treatment parameters of the as synthesized amorphous or anatase phase. Various methods have been used to prepare TiO_2 nanoparticle, such as wet chemical precipitation, microemulsion, hydrothermal crystallization and sol-gel.[9-12] Sol-gel is one of the most successful processes to prepare nanosized metal oxide materials with high photocatalytic activity and chemical purity.[13]

The aim of this work is to study the correlation between anatase-to-rutile phase ratio to photocatalytic activity of the TiO_2 powders. The effect of other properties, such as crystallite size, morphology and surface has also been investigated.

EXPERIMENTAL

Preparation of TiO_2 Photocatalyst

Tetrabutyl orthotitanate (Fluka), anhydrous ethanol (Mallinckrodt) and nitric (Scharlau) /hydrochloric (Merck) solutions were used as Ti source, solvent and reaction catalyst, respectively. Various amounts of tetrabutyl orthotitanate were dissolved in anhydrous ethanol. Then the mixture of distilled water, anhydrous ethanol and nitric solutions were slowly added and stirred at room temperature for 1-3 h. The volume ratio of H_2O: EtOH: HNO_3/HCl was varied to various compositions as summarized in Table 1. After the precursor was stirred for 1-3 h, the TiO_2 sol was then kept in a sealed container for several weeks. To obtain the TiO_2 powders, the volume of the precursor was reduced by rotor-evaporation and then dried at 50°C for 1 day. Some of the TiO_2 powders were calcined at 300-700°C for 4 h.

Table 1: Preparation conditions of various TiO_2 samples.

Sample	Type of catalyst	H_2O: EtOH: catalyst	Aging time (days)
Sol_1A	HNO_3	10:10:4	20
Sol_1B	HNO_3	10:10:5	20
Sol_2A	HCl	3:1:1	15
Sol_2B	HCl	1:3:1	30
Sol_2C	HCl	3:1:1	50

Material Characterization

The synthesized TiO_2 powders were characterized for crystalline phase by using an X-ray diffractometer (XRD; JDX-3530). The crystallite size of the TiO_2 obtained at different synthesis conditions and different heat treatments was calculated from XRD peak broadening based on Scherrer formula. The decomposition process of the residual organic compound was investigated by using a thermogravimetric analysis (TG; TGA/SDTA851). The microstructure properties were characterized by using a transmission electron microscope (TEM; JEL-2010). The specific surface area was determined from nitrogen adsorption-desorption data (Flow Sorb II 2300) based on the Brunauer, Emmett and Teller (BET) equation.

Photocatalytic Efficiency Determination

The photocatalytic efficiency of the TiO_2 was determined by methylene blue (MB) degradation under UV illumination following the method reported by Chang *et. al.*[14] Distilled water was used for the reaction solution (100 ml) containing MB (10 ppm) and 0.10 g of the TiO_2 powders in a glass container (Beaker). Then the solution was stirred vigorously for 5-10 min by using a magnetic stirrer to keep the TiO_2 particles suspended in the dispersion. Then, the aqueous dispersions were exposed to UV irradiation (24 W, λ = 253 nm). A 10 ml-aliquot was sampled at various time intervals. It was centrifuged and then filtered through a millipore membrane filter prior to analysis. Variations in the concentration of MB in each degraded solution were monitored by a UV-visible spectroscopy (V-530, Jasco Corporation, Japan)

RESULTS AND DISCUSSION

Preparation of TiO_2 Photocatalyst

Kinetic of the TiO_2 sol formation was found to be dependent on the amount of acid added into the solution, the amount of water, as well as aging time. Fig. 1 shows XRD patterns of the as-synthesized TiO_2 particles obtained from the Sol_1B, Sol_2A, Sol_2B and Sol_2C, respectively (having different ratio of H_2O: EtOH: HNO_3/HCl). According to the XRD results, the following trends were observed. Higher amount of water in the precursor led to faster sol formation, and promoted the formation of rutile phase (comparison of Sol_2B and Sol_2C). Prolong aging time also resulted in the transformation of anatase to rutile (comparison of Sol_2A and Sol_2C).

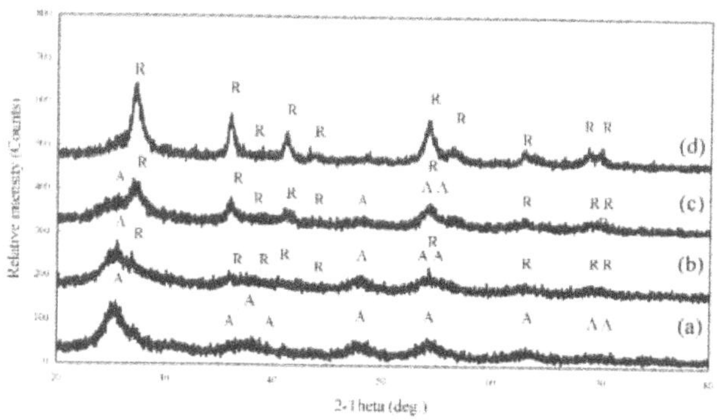

Figure 1 XRD patterns of the TiO_2 powders prepared from different conditions: (a) Sol_1A, (b) Sol_2A, (c) Sol_2B, (d) Sol_2C; A = anatase, R = rutile.

A TG curve (not shown) of the as-synthesized TiO_2 powders obtained from the Sol_1A showed two major weight losses. The first weight loss of 12.8% due to evaporation of water is observed at the temperature ranging from room temperature to 150°C. The next weight loss of 15.0% was observed at the temperature range of 150 - 300°C, which possibly due to the decomposition of an organic residues present in the as-synthesized TiO_2 powders. The thermal analysis result indicated that the synthesized TiO2 powder contained considerable amount of organic residue. A high-resolution TEM result (Fig. 2) also supported the XRD and TG analysis that the as-synthesized sample composed of the anatase crystal of average size ~ 4 - 5 nm together with an amorphous phase of either the titanium oxide or organic residue.

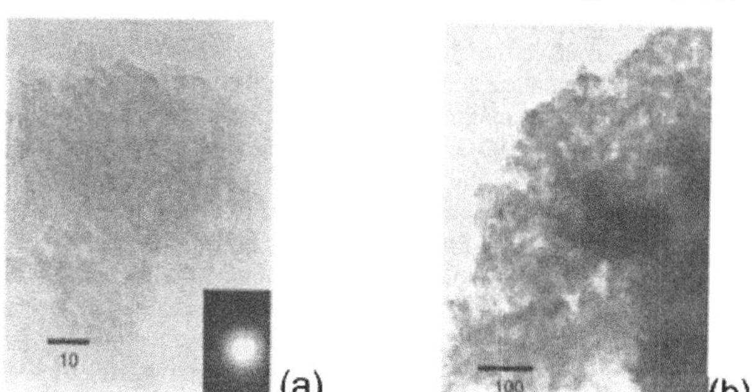

Figure 2 TEM images of the as-synthesized TiO_2 powder obtained from the Sol_1A; (a) high-resolution image, and (b) bright-field image

The effect of heat treatment on the phase composition of the TiO_2 was also studied by calcination the TiO_2 at 300, 400, 500, 600 and 700°C, respectively. The result of an XRD study is shown in Fig. 3. According to the XRD result, the as-synthesized powder from the solution condition specified contained anatase phase (Fig.3 (a)). A broad and low-intensity peak indicated that the anatase had very fine crystallite size, and may also contain an amorphous phase. The rutile structure appeared at 400°C (Fig. 3(c)). The phase ratio of the anatase to rutile at this temperature was 70:30 and decreased to 46:54 when calcined at 600°C (Fig. 3(e)). Calcination at 700°C (Fig. 3(f)) transformed all the anatase to the rutile phase.

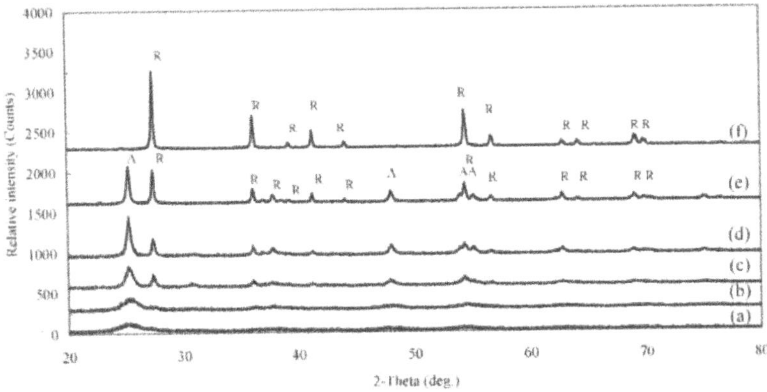

Figure 3 X-ray diffraction patterns of the TiO_2 powder synthesized from Sol_1A and calcined at different temperatures; (a) Sol_1A (as-synthesized), (b) Sol_1A3 (300°C), (c) Sol_1A4 (400°C) (d) Sol_1A5 (500°C), (e) Sol_1A6 (600°C) and (f) Sol_1A7 (700°C); A = anatase, R = rutile.

Table 2 summarizes the average crystallite size, phase composition and a specific surface area. The rutile phase composition was calculated from the equation $W_R = I_R/(0.8I_A+I_R)$ where I_A and I_R represents the integrated intensity of the anatase (110) and rutile (101) peaks, respectively.[15-16] It is obvious that increasing calcinations temperature caused the crystallite growth and anatase-to-rutile phase transformation. Fig. 4 is a TEM image of the TiO_2 powder calcined at 500 °C showing an anatase and rutile crystals of average size ~ 20 – 30 nm. Note that an amorphous phase did not exist at this heat treatment condition. The microstructure of this sample is similar to that of the P25 TiO_2.

The specific surface areas of the as-synthesized TiO_2 powders obtained from different precursor solutions were different. For instance, the specific surface area of the powder obtained from the Sol_1B was higher than that of the powder obtained from the Sol_1A. The difference of the specific surface area of the two samples could be attributed to different amount of nitric acid present in the precursor solution.[17] The TiO_2 powders prepared under high nitric acid concentration, more protons (H^+) were adsorbed on the surface of the primary particles. So most particles carried positive charge and repelled each other to form a stable sol. On the other hand, the powders prepared under low nitric acid concentration, highly aggregated clusters of the primary particles occurred. Since the powders prepared with excess nitric acid had fewer aggregates, the specific surface area was higher. It is also apparent that the specific surface area shifted towards smaller values at higher calcination temperatures since the particles aggregated. The samples obtained from the Sol_1A (uncalcined) possessed high specific surface area (239.34m^2/g), which then decreased to 38.14 m^2/g when calcined at 500°C.

Table 2 Phase composition, crystallite size and surface area of the TiO_2 powders obtained at various conditions compared to those of commercial powder, Degussa P-25.

Sample	Calcination temperature (°C)	Phase composition (A: R)	Crystallite size (nm.)		BET surface area (m^2/g)
			d_A	d_R	
Sol_1A	Noncalcined	100:0	3.87	-	167.53
Sol_1A3	300	75:25	6.39	20.19	-
Sol_1A4	400	70:30	11.61	32.55	-
Sol_1A5	500	60:40	21.23	36.47	38.14
Sol_1A6	600	46:54	30.76	48.83	-
Sol_1A7	700	0:100	-	59.76	-
Sol_1B	Noncalcined	100:0	3.39	-	239.34
Sol_2A	Noncalcined	65:35	5.50	6.89	-
Sol_2B	Noncalcined	50:50	4.10	3.73	-
Sol_2C	Noncalcined	0:100	-	9.64	-
P-25	-	85:15	22.61	30.79	49.83

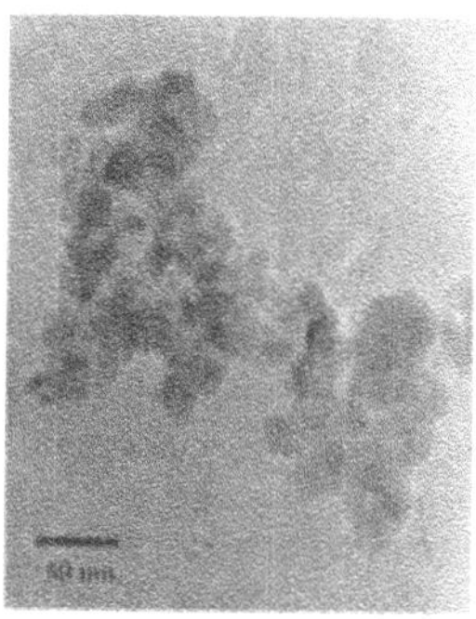

Figure 4 TEM image of the TiO_2 powder calcined at 500°C for 4 h (Sol_1A5).

Photocatalytic Activity
As-synthesized TiO_2

The complete surface adsorption of methylene blue (MB) on TiO_2 surface was ensured as the maximum absorption at 600 – 664 nm remained constant. Fig. 5 shows the results of MB decomposition of the as-synthesized TiO_2 at various times of UV irradiation. A couple of points can be drawn from this result. Firstly, the sample containing pure anatase (Sol_1A) was the most photocatalyticlly reactive. However, the activity of pure rutile was somewhat more reactive than the sample containing mixed phase. Secondly, the photoactivity of these as-synthesized powders was relatively low compared to that of the P25 whose activity result is given in Fig. 6. The present of an amorphous phase as well as an organic residue in these uncalcined powders could lower the effectiveness of an MB adsorption and decomposition of the TiO_2 crystals.

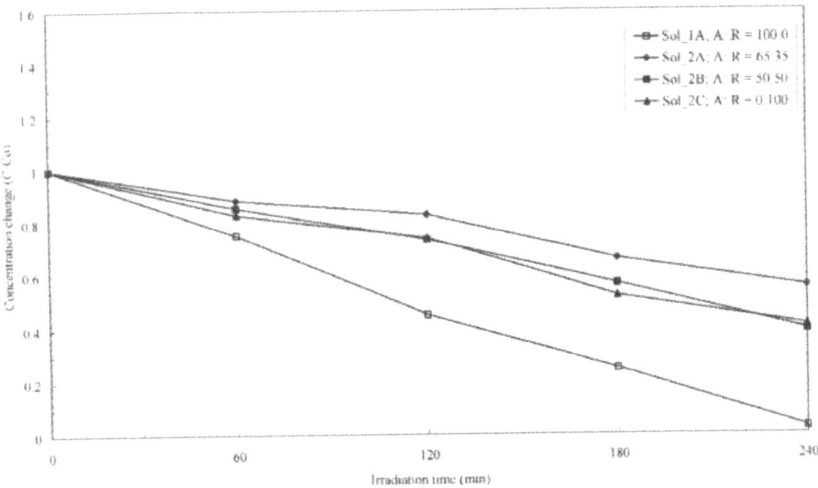

Figure 5 MB decomposition of the as-synthesized TiO_2 particles prepared at various conditions at different UV irradiation time.

Calcined TiO_2

The as-synthesized TiO_2 powders were subjected to heat treatment to eliminate an organic residue and to transform an amorphous matrix into a crystalline phase. The TiO_2 powders were found to be cleaner as no amorphous matrix was observed under TEM investigation as shown in Fig. 4 which is taken from the TiO_2 powder prepared from Sol_1A and then calcined at 500 °C (referred to as "Sol_1A5"). This sample consisted of 60 wt% anatase. When compared with the as-synthesized sample, it's MB adsorption was significantly improved. It had comparable efficiency with that of P25. Fig. 6 shows the results of MB decomposition of the TiO_2 samples obtained from the Sol_1A, and then calcined at 300 (Sol_1A3), 400 (Sol_1A4), 500 (Sol_1A5), 600 (Sol_1A6) and 700°C (Sol_1A7), respectively. Note that calcination at higher temperatures resulted to the increase amount of the rutile phase, as well as the increase of the crystallite size and the decrease of the specific surface area. According to these results, photoactivity did not directly dependent on phase composition which was varied by heating at various temperatures. For instance, the sample "Sol_1A3" which possessed higher anatase content exhibited lowest activity, probably due to the incomplete of the amorphous-anatase transformation. The amorphous TiO_2 is known to have very low photocatalytic efficiency compared to that of anatase or rutile phase due to an increased electron-hole recombination rate.[18-19] The sample "Sol_1A5" which had 60 wt% anatase exhibited the best activity which was comparable to that of a commercial P25 TiO_2.

Calcination at various temperatures not only altered phase composition, but also altered other physical properties of the TiO_2 powders, such as crystallite size, degree of crystallinity and surface area. All of these physical properties contributed to the catalytic performance. Although anatase was believed to be more photocatalytic than the rutile phase, optimum combination of surface area which usually relates to the crystallite size, cleanness of the crystal's surface and degree of crystallinity would be attributed to the best photoactivity.

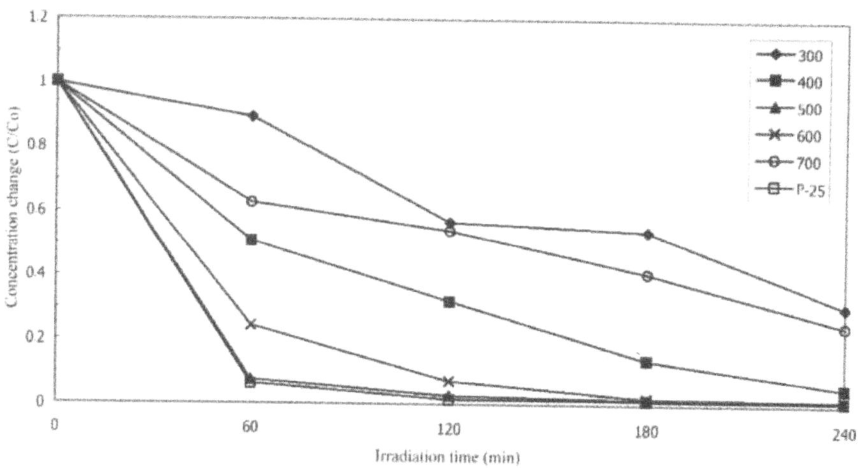

Figure 6 MB decomposition of the TiO_2 powder calcined at various temperatures.

CONCLUSION

Phase composition of the TiO_2 was varied by monitoring precursor solution during a sol-gel synthesis, as well as by heat treatment of the pure anatase at various calcination temperatures. For the as-synthesized powders, the pure anatase was the most active, while the mixed anatase and rutile phase was the least active. However, no direct correlation between a photocatalytic activity and phase composition was observed on the calcined powders since such heat treatment also alter other properties, such as crystal size, surface area, which also played important role in catalytic reaction. The highest photoactivity was observed on the sample possessed optimum values of phase composition and those physical properties.

ACKNOWLEDGEMENTS

Miss Juengsuwattananon gratefully appreciates the financial support of the Thailand Graduate Institute of Science and Technology (TGIST). The authors would like to thank the National Metal and Materials Technology Center for research grant (MT-B-47-CER-07-169-I).

REFERENCES

[1]M. Anpo, H. Yamashita, Y. Ichihashi, and S. Ehara, "Photocatalytic Reduction of CO_2 with H_2O on Variety Titanium Oxide Catalyst," *J. Electrochem.*, 396, 21-26 (1995).

[2]D. S. Bhatkhande, V.G. Pangarkar, and A.-ACM. Beenackers, "Photocatalytic Degradation for Environmental Application-A Review," *J. Chem. Technol & Biotechnol.*, 77, 102-116 (2001).

[3]A. Wold, "Photocatalytic Properties of TiO_2," *Chem. Mater.*, 5, 280-283 (1993).

[4]P.M. Sommeling, H.C. Rie!e, J.A.M. van Roosmalen, A. SchoK necker, J.M. Kroon, J.A. Wienke, and A. Hinsch, "Spectral Response and IV-Characterization of Dye-Sensitized Nanocrystalline TiO_2 Solar Cells," *Sol. Energy Mater. Sol. Cells.*, 62 399-410(2000).

[5]A.L. Linsebigler, G. Lu, J.T. Yates, and Jr, "Photocatalysis on TiO_2 Surface: Prrinciples, Mechanisms, and Selected Results," *Chem. Rev.*, 95, 735-758 (1995).

[6]S-I. Nishimoto, B. Ohtani, H. Kajiwara, and T. Kagiya, "Correlation of the Crystal Structure of Titanium Dioxide Prepared from Titanium Tetra-2-Prepoxide with the Photocatalytic Activity for Redox Reactions in Aqueous Propan-2-ol and Silver Salt Solution.," *J. Chem. Sol., Faraday Trans.*, 81, 61-68 (1985).

[7]H. Harada, and T. Ueda, "Photocatalytic Activity of Ultra-Fine Rutile in Methanol-Water Solution and Dependence of Activity on Particle Size," *Chem. Phys. Lett.*, 106, 229-231 (1984).

[8]Y. Li, T. white, and S.-H. Lim, "Structure Control and its Influence on Photoactivity and Transformation of TiO_2 Nano-Particles," *Rev. Adv. Mater. Sci.*, 5, 211-215 (2003).

[9]A. Sclafani, L. Palmisano, and M. Schiavello, "Influence of the Preparation Methods of TiO_2 on the Photocatalytic Degradation of Phenol in Aqueous Dispersion," *J. Phys. Chem.*, 94, 829-832 (1990).

[10]B. Zhang, B. Chen, K. Shi, S. He, X. Liu, Z. Du, and K. Yang, "Preparation and Characterization of Nanocrystal Grain TiO_2 Porous Microspheres," *Appl. Catal. B: Environ.*, 40, 253-258 (2003).

[11]J. Yang, S. Mei, and J.-M.F. Ferreira, "Hydrothermal Processing of Nanocrystalline Anatase Films from Tetraethylammonium Hydroxide Peptized Titania Sols," *J. European Ceram. Soc.*, 24, 335–339 (2004).

[12]C.W. Turner, "Sol-Gel Process-Principles and Applications," *Ceram. Bulletin.*, 70, 1487-1490 (1991).

[13]C. Su, B.-Y. Hong, and C.-M. Tseng, "Sol-Gel Preparation and Photocatalysis of Titanium Dioxide," *Catal. Today.*, 96, 119-126 (2004).

[14]H. Chang, C. Su, C.-H. Lo, L.-C. Chen, T.-T. Tsung, and C.-S. Jwo, "Photodecomposition and Surface Adsorption of Methylene Blue on TiO_2 Nanofluid Prepared by ASNSS," *Mater. Trans.*, 45, 3334-3337 (2004).

[15]B.D. Cullity, "Elements of X-ray Diffraction," *Addision-Wesley. USA.*, Chap. 9, 281-292 (1978)

[16]S. Music, M. Gotic, M. Ivanda, S. Popovic, A. Turkovic, R. Trojko, A Sekulic, and K. Furic, "Chemical and Microstructural Properties of TiO_2 Synthesized by Sol-Gel Procedure," *Mater. Sci. Engng. B*, 47, 33-40 (1997).

[17]Y. Jiaguo, Y. Jimmy C, L. Mitch K-P, H. Wingkei, C. Bei, Z. Xiujian, and Z. Jincai, "Effects of Acidic and Basic Hydrolysis Catalysts on the Photocatalytic Activity and Microstructures of Bimodal Mesoporous Titania," *J. Catal.*, 217, 69-78 (2003).

[18]R.I. Bickley, T. Gonzalez-Carreno, J.S. Lee, L. Palmisno, and R.J.-D. Tilleys, "A Structural Investigation of Titanium Dioxide Photocatalysts," *J. Solid State Chem.*, 92, 178-190 (1991).

[19]L. Gao, and Q. Zhang, "Effects of Amorphous Content and Particle Size on the Photocatalytic Properties of TiO_2 Nanoparticles." *Scripta. mater.*, 44, 1195-1198 (2001).

Eco-Materials Processing and Design
Edited by Koji Watari and Soo-Wohn Lee

PHOTOCATALYTIC DEGRADATION OF METHYLENE BLUE IN WATER BY UV-IRRADIATED $ATaO_3$ (A = Li, Na, K) SOL-GEL

Leticia M. Torres-Martínez, Lorena L. Garza-Tovar and Eduardo M. López R.
Ecomaterials Department, Facultad de Ingeniería Civil, Universidad Autónoma de Nuevo León,
Av. Universidad y Av. Fidel Velázquez S/N Cd. Universitaria
San Nicolás de los Garza Nuevo León, 66450 México

ABSTRACT

In this work, we report sol-gel synthesis conditions for perovskite compounds of common formula, $ATaO_3$ (A = Li, Na, K). Samples of $ATaO_3$ were prepared at pH 9 using ammonia hydroxide as hydrolysis catalyst. In addition, these materials were prepared by the traditional ceramic method in order to compare their photocatalytic effect on degradation of methylene blue. The reaction products were characterized by means of x-ray powder diffraction. The band gap (Eg) of the prepared solids was calculated by UV-Vis absorption spectra of samples annealed at different temperatures (200, 400, 600 and 800°C). The obtained values were ranged from 4.8 to 3.6 eV. Eg values of sol-gel powders of $LiTaO_3$ were higher than those of $NaTaO_3$ and $KTaO_3$. Surface characterization of the solids was carried out using adsorption isotherms and FTIR studies.

The photocatalytic activity was determined for the materials prepared by both methods on degradation of an organic dye, methylene blue, under UV radiation with λ = 254nm. The decomposition rate followed Langmuir-Hinshelwood kinetics. According to the calculated kinetic parameters, the best performance for this reaction was shown by the amorphous materials, synthesized at 400°C in the following order: $NaTaO_3 > LiTaO_3 > KTaO_3$. The presence of traces of amorphous Ta_2O_5 at this temperature resulted in the high activity of $NaTaO_3$ obtained by sol-gel method.

Keywords: Sol-gel, semiconductors, perovskites, environment, ceramic product, x-ray methods

INTRODUCTION

In the last decades, environmental photocatalytic research has been oriented to decrease and prevent pollution[1]. Heterogeneous photocatalytic reactions at semiconductor surfaces were used as a method to remove organic or inorganic pollutants from water, completely mineralizing pollutants to less harmful compounds[2-4].

The use of ultraviolet or solar light over a ceramic semiconductor material that can convert the organic compounds present in water to CO_2 and H_2O and provides the oxidation/reduction of inorganic ions into a less harmful oxidation state, (i.e. CN^- ions are oxidized to OCN^- and then to NO^{3-} and CO_2)[5,6]. Well known photocatalytic catalyst is TiO_2, with anatase structure; it is characterized by large specific surface area (40-200 m^2/g) and Eg = 3.02eV; it is activated under radiation of UV light with $\lambda \leq$ 400nm. This material is used for degradation of organic compounds such as alkanes, aromatics, herbicides, pesticides, etc.[5,6].

Degradation of phenol and 2,4-dichlorophenol has been achieved from aqueous solutions at room temperature using TiO_2 as photocatalyst and an increase of the catalytic efficiency has been possible by addition of an adsorbent like activated carbon under ultraviolet light[7,8]. These compounds could be degradated using a suspension of TiO_2 as photocatalyst, in a reactor using solar light[9]. In addition, the photocatalytic behavior of TiO_2 has been reported on methylene blue oxidation reaction, using UV light[10].

The ceramic materials with general formula ABO_3 (A= La, Y, Nd, Gd; B= Fe, Mn, Cr, Co), $LaCoO_3$ and $LaMnO_3$; which crystallize with the perovskite structure were employed for various reactions of degradation such as chlorinated volatile organic compounds (VOC's)[11,12]. For the complete oxidation of methane the following perovskite catalyst was used: $La_{0.8}Sr_{0.2}MnO_3$, $La_{0.8}Sr_{0.2}CoO_3$, $LaTi_{1-x}Cu_xO_3$, $LaNiO_3$, La_2NiO_4[13-15]. Also it was reported that $RbPb_2Nb_3O_{10}$ can be applied as photocatalyst to oxide methane under visible light (λ>420nm)[16]. It was found that $LaTi_{0.45}Cu_{0.55}O_3$[17] showed photocatatytic activity on phenol oxidation reaction. Using $BiVO_4$ it was possible to achieve the photocatalytic degradation of 4-n-nonylphenol applying visible light radiation[18]. $LaMnO_3$ supported in ZrO_2 was used for the catalytic combustion of methane[19].

On the other hand, some compounds having simple perovskite structure such as $SrTiO_3$, $ATaO_3$ (A= Li, Na, K), $AgMO_3$ (M= Ta, Nb); and layered perovskites $Sr_2M_2O_7$ (M= Nb, Ta), $K_2La_2Ti_3O_{10}$, $RbNdTa_2O_7$[20,21,22] were found to be photocatalyticaly active on water conversion into H_2 and O_2 .

The material structure is closely related with its photocatalytic activity (structure-property relationship). Wiegel et al. were the relations between crystal structure and delocalization energy[23].

On the other hand, methylene blue is an azo type dying agent, which is difficult to degradate due to its high stability at light and resistance to microbial attack. This stability of azo dyes makes them one of the most serious ecological problems since they can not be easily removed by conventional technologies, such as physical methods like adsorption and biological methods (biodegradation), and chemical methods (chlorination, ozonization)[24-27]. Also, there are some treatments for dyes degradation with chemical and biological methods that generate aromatic amines which are even more toxic than the dye itself.

In present work, the sol-gel synthesis and characterization of the perovskite family $ATaO_3$ (A = Li, Na, K) is described. The photocatalytic performance of these materials for methylene blue oxidation reaction under UV light is discussed.

EXPERIMENTAL

Solid State Reaction. Stoichiometric amounts of high purity reagents, Li_2CO_3, Na_2CO_3, K_2CO_3 and Ta_2O_5, were weighed, mixed, and pressed into pellets. Samples were then heated slowly from room temperature to 900°C for a few hours. Isothermal treatments were given from 900 to 1147°C for 1 to 10 hours depending on the composition, with intermediate regrinding to complete reaction.

Sol-Gel synthesis. Compounds were synthesized by sol-gel method using tantalum ethoxide, lithium and sodium acetylacetonate and potassium acetate. A mixture of ethanol-water was used as a solvent and reaction was carried out at acidic pH 3 by addition of acetic acid as a hydrolysis catalyst. Samples were then stirred under reflux at 70°C for 5 days, after this period pH was adjusted to 10 using NH_4OH and kept under reflux for 1 to 5 days until gel formation. Samples were cured during 24 hours and then the large amount of solvent was evaporated at 70°C. Samples were dried in an oven at 90°C in order to obtain fresh gels. For every sample different amounts were taken for heat treatments at 200, 400, 600 and 800°C for 6 hours.

Characterization.

X-ray Diffraction. Reaction products were analyzed and identified by X-ray powder diffraction using a SIEMENS D-5000 diffractometer, with $CuK_{\alpha 1}$ radiation. Simultaneous TGA/DTA was used to determine the thermal stability of sol-gel materials.

FTIR analysis. They were carried out in a Paragon 1000 (Perkin-Elmer) spectrometer (λ = 4000 to 400 cm^{-1}).

UV-Vis Spectroscopy. Diffuse reflectance measurements were carried out for each sample in the range of 900 to 300 nm, using a UV-Vis Perkin-Elmer model Lambda 12 coupled to an integration sphere RSA-PE-20. A reference standard from Spectralon USRS-99010 was used for 100% reflectance.

Textural properties. Measurements of specific surface area analysis of the samples were carried out using nitrogen adsorption isotherms with gas sorption equipment (Quantachrome Autosorb 1). The surface area was calculated by the BET method.

Photocatalytic Activity. Photocatalytic experiments were carried out at room temperature; 100mg of each catalyst was added to a flask containing 250mL of an aqueous solution of methylene blue at 30ppm. The suspension under stirring was maintained for 60 min in the dark to establish the reactant adsorption equilibrium, then the suspension was irradiated with UV light with wavelength of 254nm and an intensity of 1100 $\mu W/cm^2$, in a batch reactor inside a closed box.

RESULTS AND DISCUSSION

Solid State Synthesis of $ATaO_3$ (A = Li, Na, K)

X-ray diffraction. Reaction products obtained by solid state method were characterized by X-ray diffraction. $LiTaO_3$ and $NaTaO_3$ phases were obtained by stoichiometric mixtures of alkaline carbonates and Ta_2O_5. In order to obtain $KTaO_3$ phase an excess of 5% of potassium carbonate was added to the composition.

Sol-Gel. The sol-gel compounds $ATaO_3$ (A = Li, Na, K) were obtained under basic conditions using ammonium hydroxide as hydrolysis catalyst. X-ray diffraction powder patterns of xerogel annealed at different temperatures are shown on Figure 1, 2 and 3.

Diffraction patterns of fresh gel (70°C) and heat treated samples at 200°C and 400°C showed an amorphous pattern for three compounds. Samples heat-treated at 600°C for $LiTaO_3$ and $NaTaO_3$ show powder diffraction patterns with broad peaks evidence of the formation of crystalline phases with small size particles that correspond to patterns reported. Diffractograms of samples submitted to further heating of these materials (800°C) present well defined peaks with high level of crystallinity.

Diffraction patterns of $KTaO_3$ phase are showed in the Figure 3. As mentioned above, fresh gel and samples heat treated at 200 and 400°C presented an amorphous pattern. Sample heated at 600°C shows the formation of a crystalline phase corresponding to the tetragonal form of $KTaO_3$. For sample calcined at 800°C reflections reveals the presence of a second structure (cubic) of $KTaO_3$.

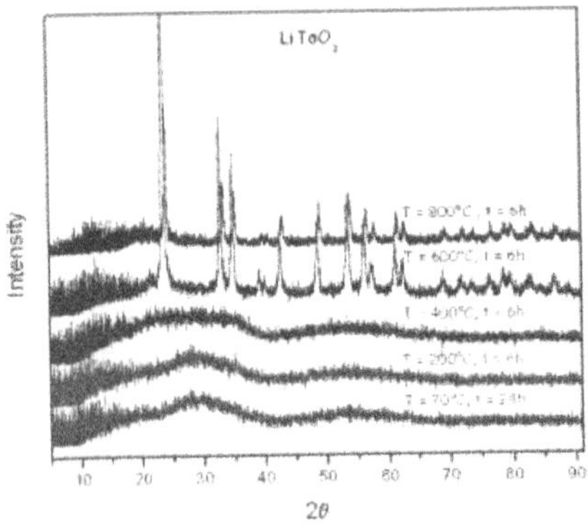

Figure 1. X-ray diffraction patterns of $LiTaO_3$ prepared by sol-gel at different temperature.

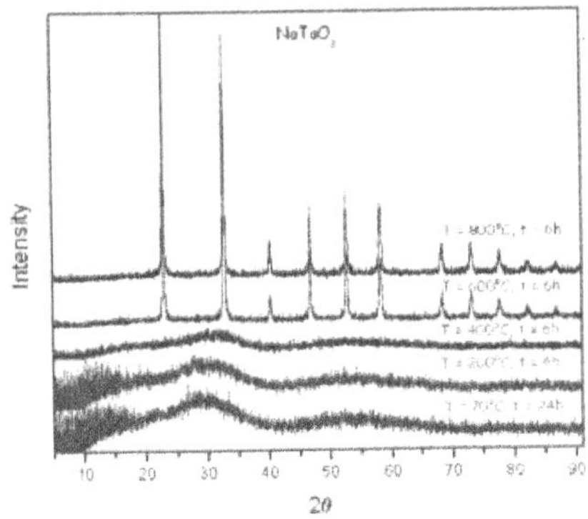

Figure 2. X-ray diffraction patterns of $NaTaO_3$ prepared by sol-gel at different temperature.

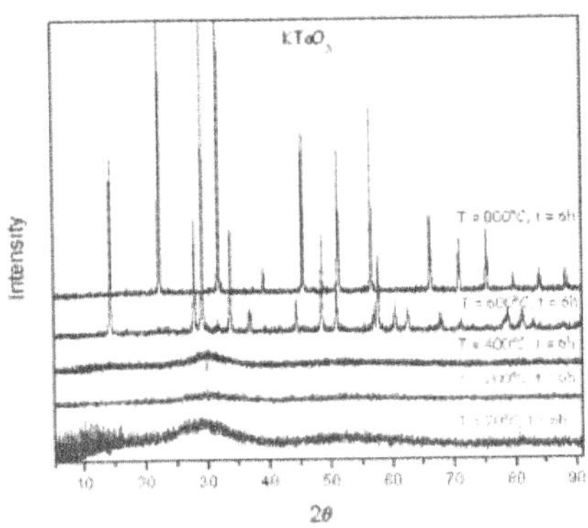

Figure 3. X-ray diffraction patterns of $KTaO_3$ prepared by sol-gel at different temperature.

Thermal Analysis

The thermogram of the fresh sample for $KTaO_3$ is presented in the Figure 4 and shows a weight loss of 22% in the temperature range 24-600°C similar to the other two family members. Four main weight losses are observed at TGA curve. First weight loss appeared between room temperature and 120°C. An endothermic peak clearly observed in this range corresponds to the lost of moisture and organic solvents. A second gradual weight loss is present in temperature range between 120 °C and 250°C and can be associated to the dehydroxilation process. Third step observed between 250 and 300°C can be associated to the decomposition of remaining organic compounds. Fourth gradual weight loss (300-500°C) according to FTIR studies can be associated with removal of some OH groups from the material.

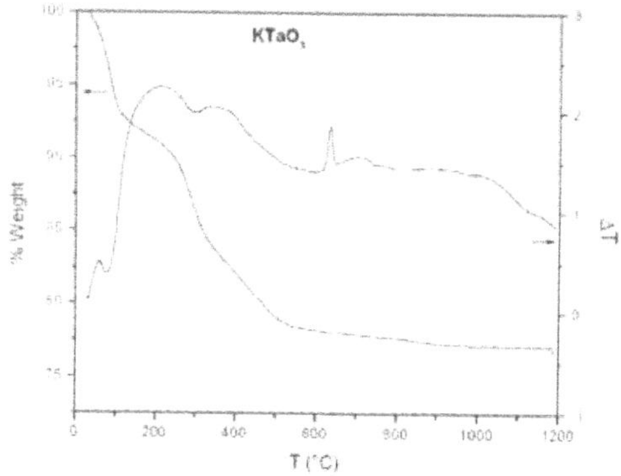

Figure 4. Thermogram of the fresh sample $KTaO_3$ prepared by the Sol-Gel method.

Approximately at 305°C there is a second endothermic peak corresponding to desorption of alkoxy groups from tantalum ethoxide and potassium acetate corresponding to beginning of the dehydroxilation of the sample. This is also observed in the FTIR spectrum where the band corresponding to the C-O bond disappears.

A clear prominent exothermic peak is observed around 630°C which indicates the crystallization of the amorphous phase. This agrees well with the results obtained by X-ray diffraction where the heat treated (at 600°C) sample shows the presence of a crystallized material. Some difference in crystallization temperature observed in X-ray diffraction and thermal analysis can be explained by the prolonged (6 h) heating in the furnace. This corresponds to the DTA results where the peak begins around 610°C.

A third endothermic peak is observed at 730°C; it indicates a structural change corresponding to that detected by the X-ray diffraction of the sample heat treated at 800°C.

FTIR Spectroscopy

Three prepared compounds demonstrated similar FTIR spectra. The infrared spectra of the fresh samples of $NaTaO_3$ are showed in the Figure 5. First, fresh sample heat treated at 70°C for 24 hours, shows wide bands around 3400cm^{-1} due to stretching vibration of O-H groups. This band decreases when the temperature increases because of the loss of water and solvent. In the sample of $KTaO_3$ heat treated at 800°C this band had almost disappeared, the remaining two samples demonstrated this band clearly, which indicates that the remaining O-H groups bond to the crystalline network. In the Figures 1, 2 and 3 (diffraction patterns) we can observe that investigated compounds are crystalline and, based on these FTIR spectra, we can say they contain small amounts of hydroxyl groups.

FTIR spectra of samples heat treated at 70°C and 200°C shows two narrow scissors type bands between 1600cm^{-1} and 1400cm^{-1}. These bands are attributed to the symmetric and asymmetric vibrations of flexion of the C-H bonds. The bands are present when acetic acid is used in the sol-gel synthesis. The bands are attributed to the functional groups CH_3 and CH_2, which are generated by organic residues such as: ethanol, acetic acid and residual alkoxides and all of these are present in the solid. These bands decrease as heat treatment temperature increases.

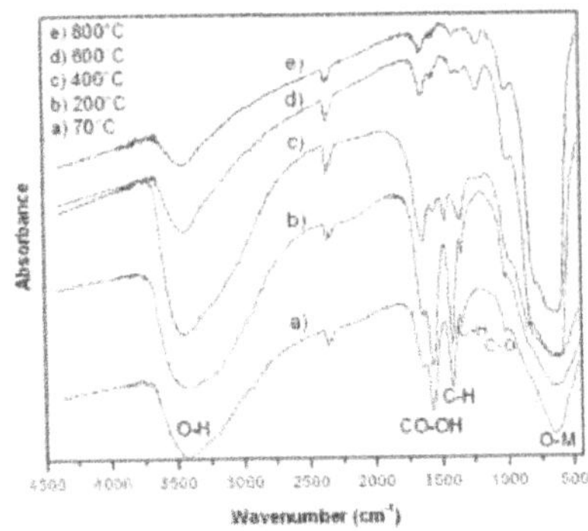

Figure 5. FTIR spectra of thermal evolution of $NaTaO_3$ sol-gel.

In addition, there is a small umbrella vibration around 1300cm^{-1}, generally indicative of the existence of organic residues associated to CH_3 groups. In the same samples, there is a small band, around 1020cm^{-1} produced by the asymmetric extension of existing C-O bonds of the residual alkoxide. These bands were not observed at 400°C.

Different heat treatments of the three compounds show a signal around 2500cm^{-1} and this is considered as a typical band that corresponds to the anti-symmetric extension of CO_2 molecule.

A typical band of the O-M bond is observed in the compounds at different heat treatments in the range from 800 to 500cm^{-1}. In the sample heat treated at 800°C, a small flexion is observed at 1000cm^{-1} which is characteristic of Ta-O bonds present in the structure of the compounds. With corresponding diffraction pattern (Fig. 1, 2 and 3) we can observe that crystalline compounds are obtained at 600°C and 800°C.

UV-VIS Spectroscopy

The energy band gap (Eg) values for the different samples, were calculated using the equation $\alpha(h\upsilon) = A(h\upsilon - Eg)^{m/2}$, where α is the absorption coefficient, $h\upsilon$ is the photon energy, A is a constant and $m = 1$ represents a direct transition between valence band and conduction band. For the estimation of Eg from the UV-Vis spectra, a straight line was extrapolated from the absorption curve to the abscissa axis. When α has a value of zero, then Eg = $h\upsilon$.

In the UV-Vis spectra of $NaTaO_3$ (Fig. 6), the sample begins to absorb UV radiation at 310nm. Also a difference between the samples heat treated at 600°C and 800°C can be observed.

The value of Eg is smaller for samples heat treated at 600°C compared with that found at 800°C. Such difference can be explained by the presence of defects and vacancies in the crystalline network of the samples heat treated at 600°C. The calculated values of Eg for the samples are reported in the table I.

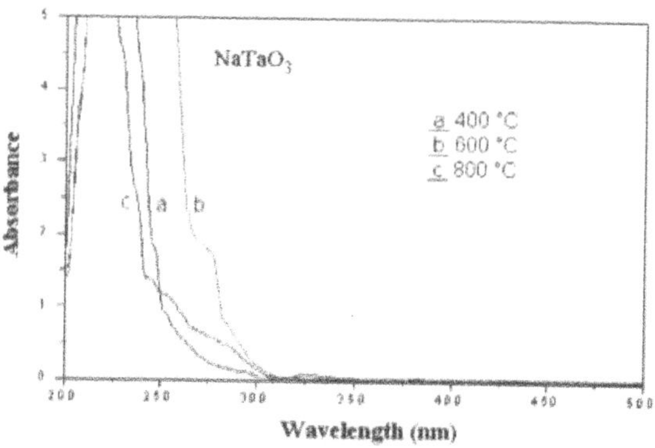

Figure 6. UV-Vis spectra of $NaTaO_3$ sol-gel heated at different temperatures.

Table I. Eg values for $ATaO_3$ (A = Li, Na and K) sintesized by sol-gel method and solid state reaction.

Compound	$LiTaO_3$ – 400°C	$LiTaO_3$ – 600°C	$LiTaO_3$ – 800°C	$LiTaO_3$-1147°C
Eg (eV)	4.45	4.69	4.80	4.63
Compound	$NaTaO_3$ – 400°C	$NaTaO_3$ – 600°C	$NaTaO_3$ – 800°C	$NaTaO_3$-1147°C
Eg (eV)	4.03	3.95	4.01	3.98
Compound	$KTaO_3$ – 400°C	$KTaO_3$ – 600°C	$KTaO_3$ – 800°C	$KTaO_3$-1147°C
Eg (eV)	4.36	4.63	3.59	3.54

Textural Properties

Figure 7 shows the nitrogen adsorption isotherms corresponding to sodium tantalate. Investigated samples heat treated at 400°C can be characterized by an isotherm of type III; this behavior is related to the adsorbent-adsorbate interactions. Low porosity of samples can be proposed from the presence of a small hysteresis of type C. The isotherms for the samples heat treated at 600 and 800°C also correspond to type III and as temperature increases, the hysteresis decrease indicating low porosity.

Table II shows the specific surface area, type of isotherm and type of hysteresis for samples calcined at 400, 600 and 800°C. Specific surface area decreases gradually as temperature increases due to the decrease on porosity of the sample.

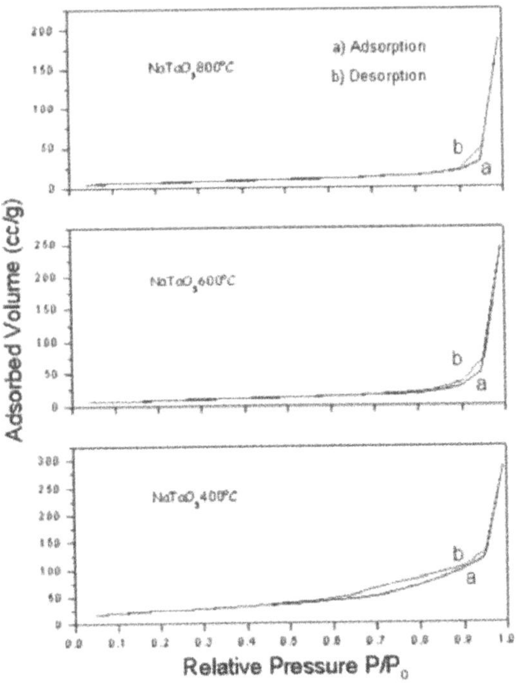

Figure 7. Nitrogen adsorption isotherms for $NaTaO_3$ solgel catalyst heated at different temperatures.

Table II. Specific surface area of $ATaO_3$ (A= Li, Na and K) sol-gel heated at different temperatures

Compound	Specific surface area (m^2/g)	Isotherm type	Hysteresis type
$LiTaO_3$ – 400°C	34.57	III	D*
$LiTaO_3$ – 600°C	15.57	III	C**
$LiTaO_3$ – 800°C	7.324	III	C
$NaTaO_3$ – 400°C	36.10	III	D
$NaTaO_3$ – 600°C	15.02	III	C
$NaTaO_3$ – 800°C	10.62	III	C
$KTaO_3$ – 400°C	5.926	III	C
$KTaO_3$ – 600°C	5.115	III	C
$KTaO_3$ – 800°C	1.895	III	C

*NNP Narrow neck porous
**IP Interconnected porous

Photocatalytic degradation of methylene blue

Photocatalytic activity of samples was determined by performing the degradation reaction of methylene blue at room temperature.

Fig. 8 shows the evolution of methylene blue concentration vs. time, using a sol-gel prepared compounds $ATaO_3$ (A = Li, Na, K) and TiO_2 (Degussa 25) as photocatalyst.

It is observed that the most efficient photocatalysts are those prepared by sol-gel method (Fig. 8), at low temperature.

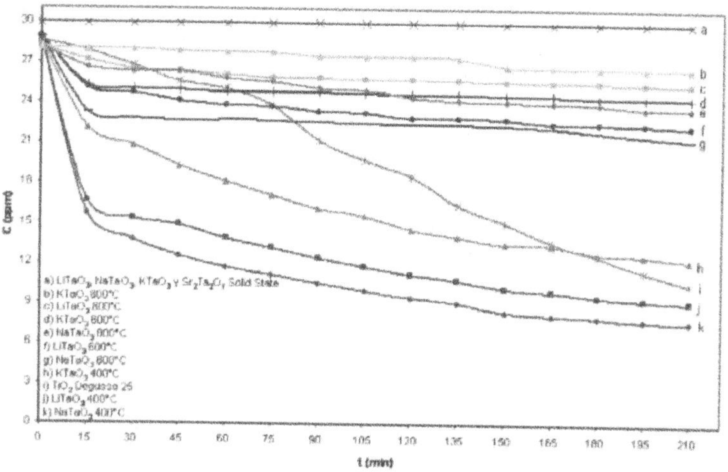

Figure 8. Photodegradation of methylene blue with $ATaO_3$ (A = Li, Na and K) photocatalysts synthesized by solid state, sol-gel method and TiO_2 Degussa P25.

In order to evaluate the photocatalytic efficiency, it was necessary to determine the kinetic parameters of the reaction, applying the Langmuir-Hinshelwood model, used for heterogeneous fotocatalytic reactions performed in batch type reactors:

$$-V\frac{dC}{dt} = \frac{mAk_1k_2C}{1+k_2C} \qquad (1)$$

where V is the liquid volume, C the substrate concentration, t is time, m catalyst mass, A the adsorption sites per gram of catalyst, k_1 the apparent rate constant and k_2 the apparent adsorption constant.

Integrating from t=0 to t=i, and for the initial concentration Co to the Ci concentration at t = i:

$$\frac{\ln\frac{Co}{C}}{Co-C} = -k_2 + \frac{mAk_1k_2}{V(Co-C)}t \qquad (2)$$

The half life ($t_{1/2}$) at C = Co / 2:

$$t_{1/2} = \frac{\left(\frac{0.693}{k_1 k_2} + \frac{Co}{2k_1}\right)}{\frac{mA}{V}} \qquad (3)$$

The kinetic parameters of methylene blue photodegradation with Langmuir-Hinshelwood model (k_1, k_2, k_1k_2 apparent reactivity and $t_{1/2}$) for catalysts obtained at different temperatures, were calculated from data of Figure 9 and are resumed in table III.

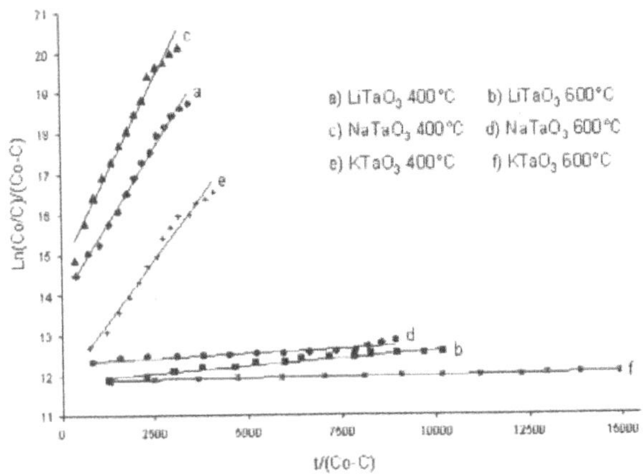

Figure 9. The kinetic paramets of Langmuir-Hinshelwood model for methylene blue degradation with catalysts $LiTaO_3$, $NaTaO_3$ and $KTaO_3$ synthesized by sol-gel

From table III, it can be seen that the best photacatalytic activity was observed for $NaTaO_3$ catalyst heat treated at 400°C, which presented a half life, $t_{1/2}$ = 21min. This catalyst also showed higher values of apparent adsorption, k_2 and hydroxylation grade (Fig 5); it can be proposed that OH groups act as an adsorption centers favoring degradation of blue methylene.

Catalytic activity of compounds calcined at 800°C had shown an abnormal behavior according to the Langmuir-Hinshelwood model. This can be attributed to small values of specific surface area of catalysts (Table II).

Table III. Parameters of the Lanmuir-Hinshelwood model for methylene blue decomposition by sol-gel catalysts.

Compound	Temperature (°C)	k_1 (min^{-1})	k_2 (M^{-1})	k_1k_2 ($min^{-1} M^{-1}$)	$t_{1/2}$	
					(min)	(h)
$LiTaO_3$	400	2.115E-04	13.90	0.00294	49	
	600	1.298E-05	11.85564	0.000154		33
$NaTaO_3$	400	2.507E-04	14.67843	0.00368	21	
	600	7.303E-06	12.32	0.00009		46
$KTaO_3$	400	2.0697E-04	11.78881	0.00244	139	
	600	2.191E-06	11.86257	0.000026		53

Therefore, applying the model for a first order kinetic in methylene blue degradation with as catalysts $LiTaO_3$, $NaTaO_3$ and $KTaO_3$ heat treated at 800°C, we have:

$$-dc/dt = kC \quad (4)$$

where k is the first order decomposition constant and C is the solute concentration.

Integrating equation 4:

$$lnC = lnC_o - kt \quad (5)$$

where C and C_o are solute concentrations at time t and 0 respectively.

Half life for reaction is given by:

$$t_{1/2} = \frac{0.693}{k} \quad (6)$$

From the data reported in Figure 10, for the degradation of methylene blue with sol-gel catalysts heat treated at 800°C the kinetic parameters were determined. Table IV shows the values of k_1 and $t_{1/2}$ for first order reaction.

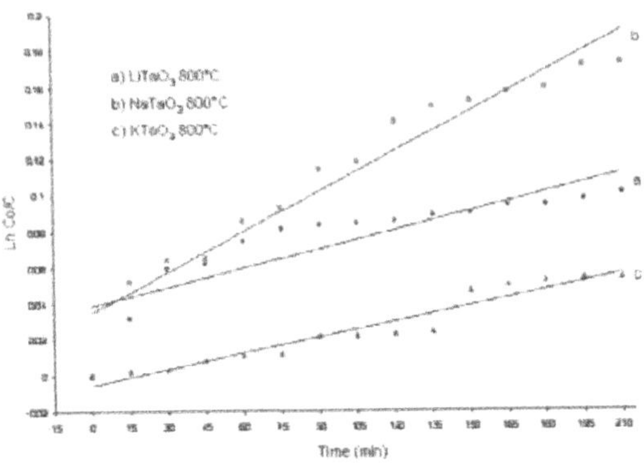

Figure 10. Photodecomposition of methylene blue with $ATaO_3$ (A = Li, Na and K) solgel calcinated at 800°C as a function of time

Table IV. Results of the first order model for methylene blue decomposition with sol-gel catalysts $ATaO_3$ (A = Li, Na and K) calcinated at 800°C

Compound	$k_1(min^{-1})$	$t_{1/2}$ (hrs)	Specific surface area (m^2/g)
$LiTaO_3$ – 800°C	3.5E-04	33.00	7.324
$NaTaO_3$ – 800°C	7.4E-04	15.60	10.62
$KTaO_3$ – 800°C	2.96E-04	39.02	1.895

According to the calculated kinetic parameters, the best performance for this reaction was shown by the amorphous materials, synthesized at 400°C in the following order: $NaTaO_3$ > $LiTaO_3$ > $KTaO_3$. X-ray routine analysis demonstrated that no crystalline phases were present in the samples heat treated at 400°C.

Figures 11 and 12 shows the comparison of long exposed x-ray powder patterns of sol-gel samples heat treated at 400 and 600°C.

For samples of $LiTaO_3$ heated at 400°C, it can be observed the concordance of the angle for the peak with maximum in 2θ between 25 and 35 and the high intensity peak of the reported pattern.

Samples of $NaTaO_3$, heat treated at 400 and 600°C show different x-ray powder patterns. From the analysis it was found that sample heated at 400°C present small peaks of Ta_2O_5. The presence of small amount of Ta_2O_5 could act as a doping agent resulting in the improvement on the photocatalytic activity of this phase.

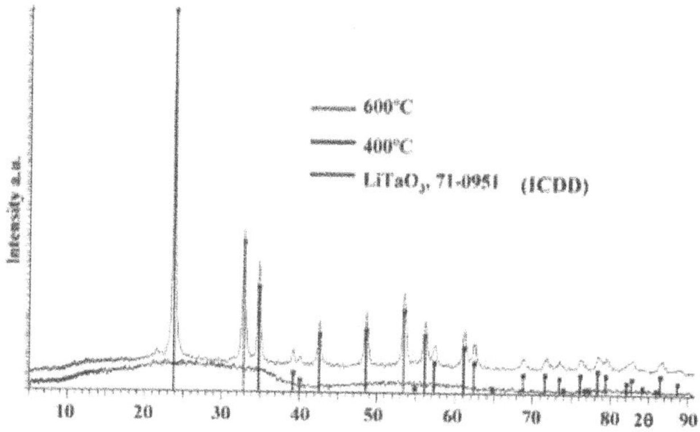

Figure 11. Comparison of X-ray powder patterns for $LiTaO_3$ sol-gel heated at 400 and 600°C

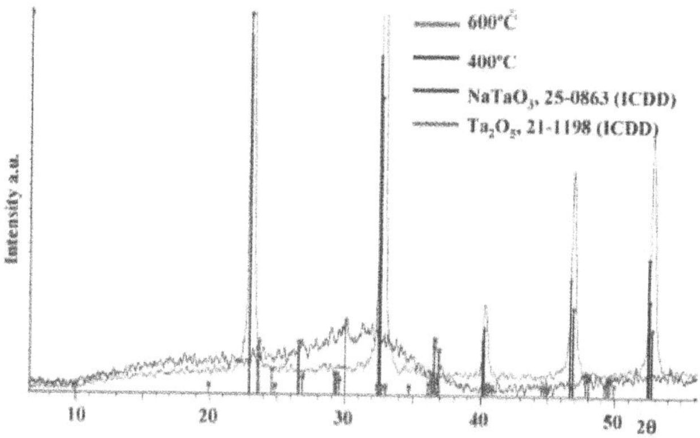

Figure 11. Comparison of X-ray powder patterns for $NaTaO_3$ sol-gel heated at 400 and 600°C

CONCLUSIONS

An alternative synthesis route for the preparation of $ATaO_3$ (A = Li, Na and K) using the sol gel method is reported. Binary phases were synthesized at lower temperatures (between 400 and 600°C) and shorter times compared to those obtained by solid-state method.

It was demonstrated that the family of compounds $ATaO_3$ (A = Li, Na and K) possess a photocatalytic activity on methylene blue reaction. Sol gel products annealed at 400°C showed better performance than those prepared by solid state reaction and TiO_2 (P-25).

According to its half life (6hrs) $NaTaO_3$ obtained by sol-gel at 400°C, possess the highest photocatalytic activity for degradation of methylene blue. It is assumed that its performance could be attributed to the presence of a second phase (Ta_2O_5) that may act as a dopant, improving the electronic transfer between the tantalum atoms present in this phase and those present in MO_6 octahedral of the perovskite structure.

Materials of the family $ATaO_3$ (A = Li, Na and K) synthesized by sol-gel method and annealed at 800°C, showed the lowest photocatalytic activity in the methylene blue oxidation reaction, possibly because of their high degree of sintering, lower porosity and absence of hydroxyl groups in the structure .

According to the results obtained, band gap energy is a determinant factor in the methylene blue degradation, since photocatalytic efficiency decreases as Eg values increases.

ACKNOWLEDGEMENTS

Authors would like to express their gratitude to CONACYT (Grants 35415-U), SEP-CONACYT (Grant 42910-R), SEMARNAT (Grant 2004-01-C01-00394) and UANL (PAYCTY-2005) for their financial support of this research program.

REFERENCES

[1]Farrauto R. J., Heck R. M.; Environmental catalysis into the 21st Century, Catalysis Today, (2000), 55, 179-187.

[2]Hermann J.M., Heterogeneous photocatalysis: Concepts, reaction mechanisms and potential applications in environmental problems, Trends in Photochemistry & Photobiology, (1994), 3, 633-642.

[3]Hermann J.M., Matos J., Disdier J., Guillard C., Laine J., Malato S., Blanco J.; Solar photocatalytic degradation of 4-chlorophenol using the synergistic effect between titania and activated carbon in aqueous suspension, Catalysis Today, (1999), 54, 255-265.

[4]Vidal A., Díaz A.I., El Hraiki A., Romero M., Muguruza I., Senhaji F., González J.; Solar photocatalysis for detoxification and disinfection of contaminated water: pilot plant studies, Catalysis Today, (1999), 54, 283-290.

[5]Hermann J.M., Heterogeneous photocatalysis: an emerging discipline involving multiphase systems, Catalysis Today, (1995), 24, 157-164.

[6]Hermann J.M., Heterogeneous photocatalysis: fundamentals and applications to the removal of various types of aqueous pollutants, Catalysis Today, (1999), 53, 115-129.

[7]Matos J., Laine J., Hermann J.M.; Effect of the type of activated carbons on the photocatalytic degradation of aqueous organic pollutants by UV-irradiated titania, Journal of Catalysis, (2001), 200, 10-20.

[8]Scalfani A., Palmisano L., schiavello M.; Influence of the preparation methods of TiO_2 on the photocatalytic degradation of phenol in aqueous dispersion, J. Phys. Chem., (1990), 94, 829-832.

[9]Giménez J., Curcó D., Queral M.A.; Photocatalytic treatment of phenol and 2,4-dichlorophenol in a solar plant in the way to scaling-up, Catalysis Today, (1999), 54, 229-243.

[10] Hinda Lachheb, Eric Puzenat, Ammar Houas, Mohamed Ksibi,Elimame Elaloui, Chantal Guillard, Jean-Marie Herrmann ; Photocatalytic degradation of various types of Dyes (Alizarin S, Crocein Orange G, Methyl Red, Congo Red, Methylene) in water by UV-irradiated titania.

[11]Poplawski K., Lichtenberger J., Keil F. J.; Schnitzlein, K.., Amiridis, M. D.; Catalytic oxidation of 1,2-dichlorobenzene over ABO_3-type perovskites, Catalysis Today (2000), 62(4), 329-336.

[12]Sinquin G., Petit C., Hindermann J. P., Kiennemann A.; Study of the formation of $LaMO_3$ (M = Co, Mn) perovskites by propionates precursors: application to the catalytic destruction of chlorinated VOCs., Catalysis Today (2001), 70(1-3), 183-196.

[13]Xue Ping; Shen, Yuenian; Sun, Yanhua; Hu, Ruisheng..; Studies on the support effect of $La_{0.8}Sr_{0.2}MnO_3$ and $La_{0.8}Sr_{0.2}CoO_3$ loading on α-Al_2O_3, Fenzi Cuihua (1998), 12(6), 424-428.

[14]Van Grieke, R.;, Pena J. L., Lucas A., Calleja G., Rojas M. L., Fierro, J. L. G.; Selective production of methanol from syngas over lanthanum titanium copper oxide ($LaTi_{1-x}Cu_xO_3$) mixed oxides, Catal. Lett. (1991), 8(5-6), 335-44.

[15]Ladavos, A. K.; Pomonis, P. J.; Skaribas, S. P., Catalytic activity of perovskite species $LaNiO_x$ intercalated into montmorillonite as compared to nonintercalated ones, Mater. Sci. Forum, (1992), 91-93, 799-804.

[16]Yoshimura J., Ebina Y., Kondo J., Domen K., Tanaka A.; Visible light-induced photocatalytic behavior of a layered perovskite-type rubidium lead niobate, $RbPb_2Nb_3O_{10}$. J. Phys. Chem. 1993), 97(9), 1970-3.

[17]Ovejero G., Sotelo J. L., Martinez, F., Gordo, L.; Novel heterogeneous catalysts in the wet peroxide oxidation of phenol, Water Science and Technology (2001), 44(5, Oxidation Technologies for Water and Wastewater Treatment II), 153-160.

[18]Kohtani S., Makino S., Kudo A., Tokumura K., Ishigaki Y., Matsunaga T., Nikaido O., Hayakawa K., Nakagaki R.; Photcatalytic degradation of 4-n-nonylphenol under irradiation from solar simulator: comparision between $BiVO_4$ and TiO_2 photocatalysts, Chemistry Letters, (2002), 7, 660-661.

[19]Cimino S., Pirone R., Lisi L.; Zirconia supported $LaMnO_3$ monoliths for the catalytic combustion of methane, Applied Catalysis, B: Environmental (2002), 35(4), 243-254.

[20]Kudo A., Kato H., Nakagawa S.; Water Splitting into H_2 and O_2 on New $Sr_2M_2O_7$ (M=Nb and Ta) Photocatalysts with Layered Perovskite Strucutres: Factors Affecting the Photocatalytic Activity, American Chemical Society J. Phys. Chem. B., (2000), 104, 571-575.

[21]Kato H., Kudo A.; Water Splitting into H_2 and O_2 on Alkali Tantalate Photocatalysts $ATaO_3$ (A= Li, Na and K), J. Phys. Chem. B., (2001), 104, 571-575.

[22]Kato H., Kobayashi H., Kudo A.; Role of Ag^+ in the band structures and photocatalytic properties of $AgMO_3$ (M= Ta and Nb) with the perovskite structure, J. Phys. Chem. B., (2002), A-G.

[23]Wiegel M., Emond M. H. J., Stobbe E.R., Blasse G.; J. Phys. Solids, (1994), 55, 773.

[24](a) P.B. Dejohn, R.A.Hutchins, Text. Chem. Color. 8 (1976) 69; (b) A. Houas, I. Bakir, M. Ksibi, E. Elaloui, J. Chim. Phys. 96 (1999) 479.

[25]S.S. Patil, V.M. Shinde, Environ. Sci. Technol. 23 (1988) 1160.

[26]A.T. More, A. Vira, S. Fogel, Environ. Sci. Technol. 23 (1989) 403.

[27] Y.M. Slokar, A.M. Le Marechal, Dyes Pigments 37 (1998) 335.

Eco-Materials Processing and Design
Edited by Koji Watari and Soo-Wohn Lee

APPLICATION OF COMPUTATIONAL CERAMIC ENGINEERING TO PHOTOCATALYTIC TiO_2, DEFECT STRUCTURES IN TiO_2, AND MULTIDOMAIN FERROELECTRICS

J. Brandon Keith, H. Wang, Dorian M. Hatch, and James P. Lewis
Department of Physics and Astronomy
Brigham Young University
Provo, UT, 84604

ABSTRACT

Computational tools on all time and length scales are playing an increasingly important part in ceramics development and manufacturing. As a survey of recent advances in the micro- and mesoscales this work explores bandgap and defect engineering of TiO_2 and domain average engineering of ferroelectrics. Specifically, anion doping in TiO_2 with nitrogen, phosphorous, carbon and sulfur is emphasized in this work and a discussion of the electronic structure and localization for each of these systems is presented. It is found sulfur, phosphorous and low-concentration carbon and nitrogen are fair to poor photocatalytic dopants and high-concentration carbon and nitrogen are better photocatalysts. The effect of oxygen vacancy defects in TiO_2 and the free energetic barrier of these defects are discussed. A method to calculate the free energy of formation of oxygen vacancies using DFT local orbital techniques and initial results thereof are presented. The mesoscopic regime of multidomain configurations in ferroelectrics is investigated. A brief synopsis of a computationally efficient real space model for ferroelectrics is given and illustrated using domain average engineering of perovskite structures guided by multidomain group theory.

INTRODUCTION

Ceramics are some of the most useful materials known to man and efforts to profitably employ them in fields such as catalysis and actuators have undergone extensive development. With the advent of increased computational power has also come the ability to refine their composition, structure, and processing by means of advanced computational models. This work surveys some of these advances in three technologically important areas of ceramics development: band gap engineering, defect engineering, and domain average engineering. These three applications illustrate the variety of time and length scales inherent in ceramics modeling and some of the challenges to integrate modeling in ceramic design and manufacturing.

The smallest characteristic length belongs to electronic structure calculations. Current issues in ceramic electronic structure include how to deal with large supercells required when treating ceramic impurities and defects. This issue is examined through a study on TiO_2 anion doping for photocatalysis. Another phenomenon, defects, also require large supercells to model but additionally involve long time scales to calculate temperature dependent concentration using quantum MD. In both of these areas a fast yet accurate ab initio method must be applied. FIREBALL[1] has been chosen to meet that need in this work. FIREBALL is a tight-binding molecular dynamics (TBMD) code based on a self-consistent version of the Harris-Foulkes functional. Confined atomic-like orbitals form a basis set for the determination of the occupied eigenvalues and eigenvectors of the one-electron Hamiltonian. The ``fireball" orbitals, introduced by Sankey and Niklewski[2], are obtained by solving the atomic problem with the boundary condition that the atomic orbitals vanish outside and at a predetermined radius where wavefunctions are set to zero.

An important advantage of the Sankey and Niklewski basis set is that the Hamiltonian and the overlap matrix elements of the system are quite sparse for large systems, reducing overall computation time. FIREBALL has been successfully applied to a number of systems such as zeolites, clathrate structures, semiconducting materials, and biomolecules[1,3,4,5].

As one approaches the mesoscale regime in ceramics modeling, atom-based methods become too expensive but macroscopic methods do not sufficiently resolve microstructure. Examples include the multidomain structure typically formed during ceramic processing in ferroelectrics cooled to below the Curie point. This topic is investigated here using phase field theory, which is based on fundamental principles of thermodynamics and kinetics and provides a powerful method for predicting the temporal evolution of microstructures. This is done by solving kinetic equations of spatially inhomogeneous order parameters without any prior assumptions of transient microstructure. This work examines perovskite structures and uses group theoretical guidance to simulate novel multidomain configurations.

The outline of the remainder of this paper is as follows. First bulk properties of TiO_2 are calculated and nitrogen, phosphorous, sulfur, and carbon doped TiO_2 band structure and wavefunction localization are calculated. Next oxygen vacancies in TiO_2 are discussed and a method to calculate their free energies of formation is presented. Preliminary results are set forth. Ferroelectric multidomain structures are then discussed in the context of domain average engineering and a group theoretical derivation of all possible multidomain structures is briefly reviewed. Group theory is used to guide multidomain processing of $BaTiO_3$-type materials. A discussion of these techniques and the results achieved in these systems is presented.

CERAMIC SYSTEMS STUDIED AND RESULTS OBTAINED

TiO_2 Photocatalysis

TiO_2 can be used as a photocatalyst for splitting water, self cleaning of tile, glass, and windows, air purification, deodorization, and sterilization. With high chemical stability, nontoxicity, and a relatively low-price, TiO_2's only serious disadvantage is that only UV light (3% of the solar spectrum) can be used to activate it because of its wide intrinsic band gap. It is of great interest to find ways to engineer the absorption wavelength range of TiO_2 into the visible region without a decrease in photocatalytic activity.

Impurity doping is frequently used to improve optical absorption. Cation doping generally decreases optical absorption even in the UV region due to an increase in the number of recombination centers produced by localized d-states deep in the band. Anion nonmetal dopants, such as carbon, nitrogen, sulfur, or phosphorous may be more appropriate for the extension of photocatalytic activity into the visible-light region because impurity states are near the valence band edge but do not act as charge carriers. Furthermore, their role as recombination centers may be minimized compared to cations. Experiments show a redshift in the absorption spectra to the visible region. In this work phosphorous doping is studied and nitrogen, sulfur, and carbon doping of TiO_2 are reviewed.

Our doping calculations are benchmarked against electronic/optical properties of pure TiO_2. FIREBALL gives predictions of the tetragonal lattice constants a and c and internal parameter u of rutile and anatase TiO_2 within one percent[6] of experimental measurements[7] and a direct band gap of 3.05 eV and 3.26 eV[8] at the Gamma point for rutile and anatase polymorphs respectively. These predictions are in excellent agreement with the experimental gaps 3.06 eV[9] and 3.20 eV[10].

Nitrogen Doping

Experimental results for doping with nitrogen indicate there is a significant red shift in the absorption spectra[11] which yields photoactivity in the visible. To explore the effects of nitrogen doping[8] a 384 atom supercell of 5.2% and 0.52% substitutional N doped rutile was used to calculate the electronic density of states (DOS) shown in Fig. 1. The top panel shows the DOS for bulk TiO_2 and the lower panels N doped TiO_2 and the indicated projections. The dashed line indicates the top of the valence band. It is apparent there is no valence band shift for higher N concentrations but a significant shift at lower N concentrations with a bandgap of 2.55 eV, in agreement with experimental observations[11].

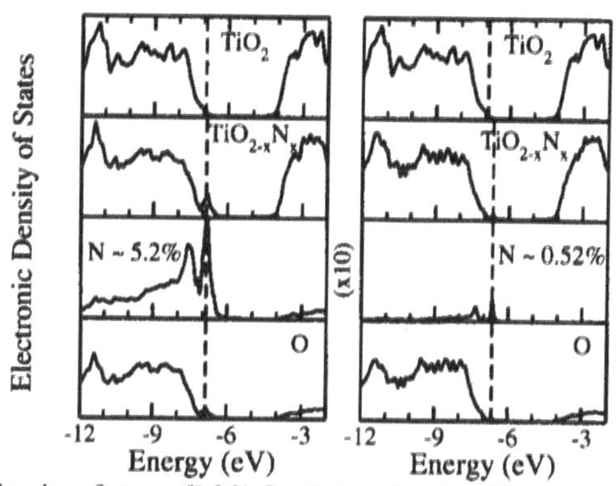

Figure 1: Electronic density of states (DOS) for N doped rutile TiO_2 at 5.2% (left) and 0.52% (right) concentrations. The dashed line indicates the valence band edge in the system. The top panel shows the DOS for bulk TiO_2 as a reference, the second panel shows the N doped structure, and the third and lowest panels show the projected DOS onto nitrogen and oxygen atoms, respectively.

To calculate the degree of electron localization near the band edge a parameter W, which gives the number of accessible atoms in a particular electronic state[12], is used. Figure 2 shows a scatterplot of W for each electronic state near the valence band edge for 5.20% and 0.52% N doped anatase TiO_2. The dashed line indicates the valence band edge. It has been proposed[11] that the states in the gap should overlap sufficiently with the O_{2p} states to transfer photoexcited carriers to reactive sites at the catalyst surface within their lifetime. The projected DOS onto nitrogen and oxygen atoms (Fig. 1, left) shows there is significant overlap between the nitrogen states and O_{2p} states for 5.20% N doped TiO_2. Figure 2 shows these states are highly delocalized, allowing electrons to reach the surface more easily. On the other hand at lower concentrations the states introduced by N dopants are distinct and highly localized on a single N atom. There is no significant overlap observed for the structures of 0.26% N concentration. And the energy shift of the valence band edge is much smaller compared with the high doping concentration structures. This allows one to predict that photocatalytic performance for smaller N concentrations should be much lower than that of higher concentrations. Calculations for anatase nitrogen doped TiO_2 show similar trends except there is a smaller shift in the band gap (2.82 eV) at both higher and lower nitrogen concentrations and roughly equivalent types of localization at both concentrations[8].

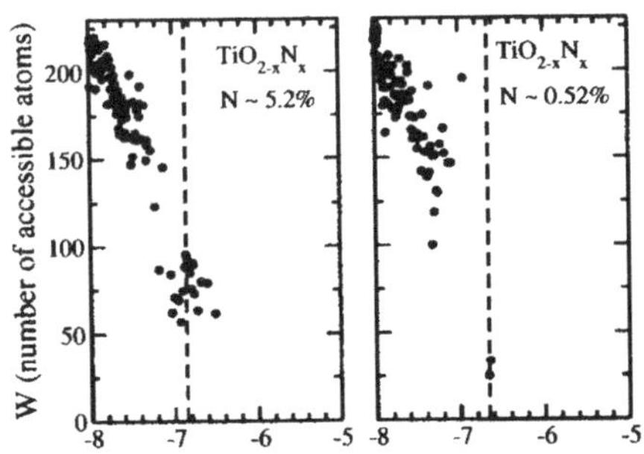

Figure 2: Accessible atoms per electronic state for N doped TiO_2. 5.2% and .52% doping concentrations are shown on the left and right. The dashed line is the valence band edge.

Phosphorous Doping

Experimental results of phosphorous doping in anatase TiO_2 have recently been reported[13]. To explore the effects of P doping the same 384 atom supercell of 5.2% and 0.52% substitutional P doped anatase is used to calculate the electronic density of states (DOS) shown in Fig. 3. The top panel shows the DOS for bulk TiO_2 and the lower panels P doped TiO_2 and the indicated projections. The dashed line indicates the top of the valence band. It is apparent that phosphorous doping induces a very large change in the Fermi level, shifting the bandgap from 3.26 eV to 1.541 eV and 1.601 eV in the larger and smaller concentrations. This is in qualitative agreement with preliminary experimental results for 3.4% doped anatase, showing a redshift in the absorption spectrum but without such a large bandgap[13]. Also apparent in the third and fourth panels is the smaller overlap of the phosphorous p states with the O p states which signifies a lower mobility of photoexcited carriers than in N doped TiO_2.

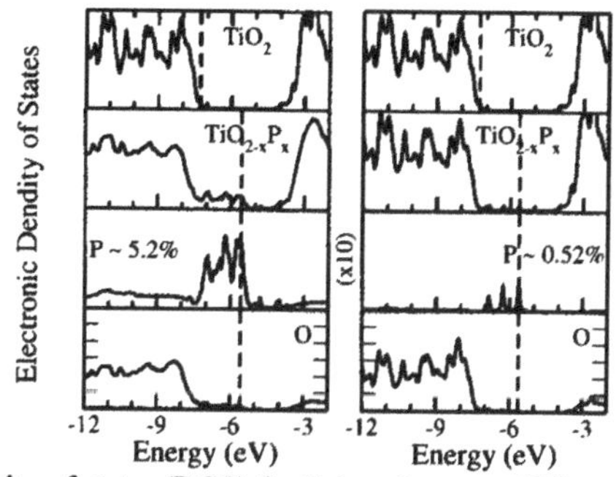

Figure 3: Electronic density of states (DOS) for P doped anatase TiO_2 at 5.2% (left) and 0.52% (right) concentrations. The dashed line indicates the valence band edge in the system. The top panel shows the DOS for bulk TiO_2 as a reference, the second panel shows the P doped structure, and the third and lowest panels show the projected DOS onto P and O atoms, respectively.

The wave function localization of P doped anatase is shown in Fig. 4 for the two concentrations mentioned before. It is apparent that at low concentrations the valence band edge states are highly localized and are not sufficiently extended to transfer photoexcited carriers to the surface, whereas at higher concentrations there is a moderate degree of delocalization and therefore a better likelihood of efficient photocatalysis. It seems this partial localization and lack of mixing with O p states may be responsible for P doped TiO_2's poor performance as a photocatalyst based on solely expectations from band gap shifts. Unoptimized sample preparation may also have contributed[13].

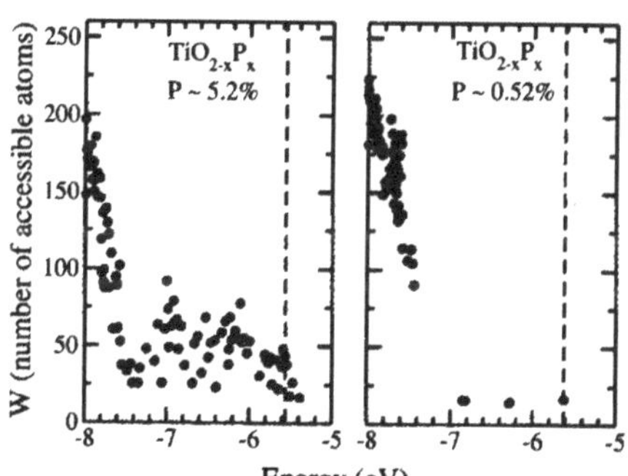

Figure 4: Accessible atoms per electronic state for P doped TiO_2. 5.2% and .52% doping concentrations are shown on the left and right. The dashed line is the valence band edge.

Other Types of Anion Doping

Sulfur and carbon doping are now briefly reviewed. First, although some theoretical work shows that S doping contributes to band gap narrowing, experimental realization was difficult until recently in the case of anatase polycrystalline powder[11]. A DOS analysis[8] (not shown) for 5.2% and 0.26% sulfur doped anatase showed no significant valence band shift at low concentrations but at 5.2% concentration the band gap is reduced to 2.2 eV, in agreement with other sources[14]. However, a calculation of the number of accessible atoms for electronic states near the valence band edge[8] showed little delocalization, impeding the mobility of charge carriers and greatly decreasing its photocatalytic activity overall.

The DOS calculation for 5.2% and 0.26% C doped rutile TiO_2 (not shown) by Wang and Lewis[15] reveals bandgap narrowing due to new half-filled carbon 2p states above the valence band edge. In the 5.2% doped structure there is a significant shift in the valence band edge, resulting in a narrowed band gap of 2.35 eV. Other studies confirm a red shift in the absorption spectra at this concentration[11]. At 0.26% C doped rutile the band is shifted by about the same amount but little overlap with oxygen valence band states occurs. A calculation of the number of accessible atoms per electronic state (not shown) at 0.26% C doping reveals the valence band edge states are also quite localized. As a result hole mobility will be limited, leading to reduced photocatalytic activity. For 5.2% C doped TiO_2 the states near the valence band edge are delocalized and C states overlap significantly with oxygen atom states. This explains why higher carbon concentration produces visible absorption with larger photocatalytic efficiency overall.

Vacancy Defects in TiO_2

Another factor affecting TiO_2's optical and electronic properties is its defect state[16]. Intrinsic defects such as oxygen vacancies result in n-type doping and high conductivity[17]. Moreover, such defect concentrations vary greatly with processing and annealing temperature[17]. Calculations to capture such effects must use *ab initio* thermodynamics, an increasingly important component of computational ceramic engineering. This work looks at formation free energies of oxygen vacancies, the most common TiO_2 defect structure and one that has a dominating effect on surface chemistry[18]. Again FIREBALL is used because it is accurate and fast enough to handle *ab initio* thermodynamic ensembles. It is also appropriate since empirical potentials typically are not constructed and benchmarked to take into account the large electron density variations near vacancies.

To calculate vacancy formation free energy, a nonequilibrium thermodynamic integration scheme is used (TI). This is described in detail in Ref. 19 although a short synopsis is given here. TI uses a coupling parameter λ between the Hamiltonians H of multiple systems, in this case a perfect crystal (pc), defect crystal (dc), and oxygen vacancy (o),

$$H = (1-\lambda)H_{pc} + \lambda H_{dc} + \lambda H_{o}. \tag{1}$$

The derivative of the Helmholtz free energy can be shown[20] to be the thermal average of the λ derivative of the potential at constant λ,

$$\frac{\partial F(\lambda')}{\partial \lambda'} = \left\langle \frac{\partial V(r_i,\lambda)}{\partial \lambda} \right\rangle_{\lambda'}, \tag{2}$$

which gives the reversible work and free energy difference when integrating from the perfect crystal to the defect one,

$$\Delta F = W_{rev} = \int_{\lambda'=0}^{\lambda'=1} \left\langle \frac{\partial V(r_i,\lambda)}{\partial \lambda} \right\rangle_{\lambda'} d\lambda' . \tag{3}$$

Typically λ is discretized and a series of equilibrium simulations are done to find this difference. A nonequilibrium TI simulation dynamically varies the coupling parameter during the MD simulation to give the irreversible work of vacancy formation. This integration is typically done twice to achieve upper and lower bounds to the free energy difference. An average of the two runs at appropriate simulation length gives the reversible work and free energy difference[21].

This reversible work of free energy formation for an oxygen vacancy in TiO_2 over a limited temperature range is shown in Fig. 5. Assuming monovacancies are the primary contributor to the defect population, the defect concentration may be calculated to a good approximation using an Arrhenius type expression, $c = e^{-F_f/kT}$, for concentration c[22].

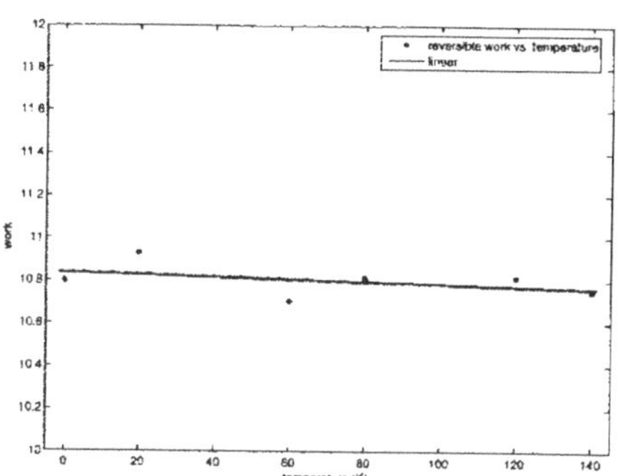

Figure 5: Preliminary results for the reversible work of oxygen vacancy formation in TiO_2 as a function of temperature calculated using *ab initio* nonequilibrium thermodynamic integration.

FERROELECTRIC MICROSTRUCTURE

Defects such as vacancies are particularly concentrated near crystal structure boundaries in polycrystals or at domain boundaries which are common in ferroelectrics cooled to below the Curie point. Examples include barium titanate, lead titanate, potassium niobate, etc. Similar to defects, the presence of multiple domains has a large impact on material properties, such as dielectric and piezoelectric constants. Control over how such microstructure forms would lead to fine-grained control over these important material properties and the ability to enhance them, such as increasing the piezoelectric strain under external electric fields.

Previous work[23,24,25] has shown the ability to simulate ferroelectric microstructure, mostly in the case of 2D lattices. This work considers domain engineering for a 3D lattice. Domain engineering generally deals with two kinds of structural descriptions, domain geometric engineering and domain average engineering. Domain geometric engineering attempts to alter the geometrical configurations of individual domains, including twinning patterns, domain wall orientation, and positioning. Domain average engineering deals with an average of domain configurations without considering the details of spatial occupation. Allowable symmetry operations on domain average structures refer to domain volume fractions and may not bring the local structure back to itself. However, they do leave the domain volume ratios intact.

This work considers the second description of microstructure and so deals with domain average engineering (DAE), which is outlined in detail in Ref. 26. DAE uses external fields such as deviatoric and shear strains (Γ_3^- and Γ_5^- space group irreducible representations (IRs) in space group 221) or electric fields (Γ_4^+ space group IR) to shift the original volume fractions of domains to one more appropriate for a given application. Table I shows an example of DAE for the 221 to 99 phase transition as found in barium titanate or lead titanate when the unit cell elongates in the [100] direction or any of the six equivalent directions. The 'Domain Set' column shows six 'a' constants in the first row representing the six equal volume fractions of the first [100] domain, the second $[\bar{1}00]$ domain, the third [001] domain, and so on. The 'Group' column shows the point group which operates on the crystal and leaves the domain volume fractions intact and is discussed in detail in Ref. 26. The space group IRs in the third, fourth, and fifth columns are the type of external field that must be applied to get new ratios of volume fractions as shown below the first row in the 'Domain Set' column.

Table I: Some symmetrically distinct multidomain configurations for ferroelectric ordering along [100].

Domain Set	Group	Γ_4^-	Γ_3^+	Γ_5^+
(a,a,a,a,a,a)	$m\bar{3}m$			
(a,a,b,b,a,a)	$4_z/m_x m_{xy} m_z$		(a,0)	
(a,b,a,b,a,b)	$3_{xyz} m_{\bar{x}y}$	(a,a,a)		(a,a,a)
(a,a,b,a,a,a)	$4_z m_x m_{xy}$	(0,0,a)	(a,0)	
(a,a,b,c,c,b)	$m_{yx} m_x 2_{\bar{y}z}$	(0,a,-a)	$(a,a\sqrt{3})$	(0,a,0)

The phase field model discussed in the introducion which is appropriate to demonstrate DAE is described in detail in Ref. 27, so only a brief synopsis is given here. It consists of a free energy functional over polarization $\vec{P}(\vec{r})$ and strain displacements $\vec{u}(\vec{r})$,

$$F[\vec{P}(\vec{r}),\vec{u}(\vec{r})] = \int(f_P + f_u + f_c + f_E + f_{dep})d\vec{r}\,, \tag{4}$$

with contributions from polarization and its gradient f_P, strain displacements f_u, a coupling term between polarization and strain f_c, an electric field interaction term f_E, and a depolarization term f_{dep} that ensures an approximate overall unpolarized state. This free energy functional is evolved in time using Langevin dynamics to achieve a multidomain configuration as shown on the left in Fig. 6 where the six distinct domains of the 221 to 99 phase transition are represented by six distinct colors. Their approximately equal volume fractions are shown below on the left in a bar chart.

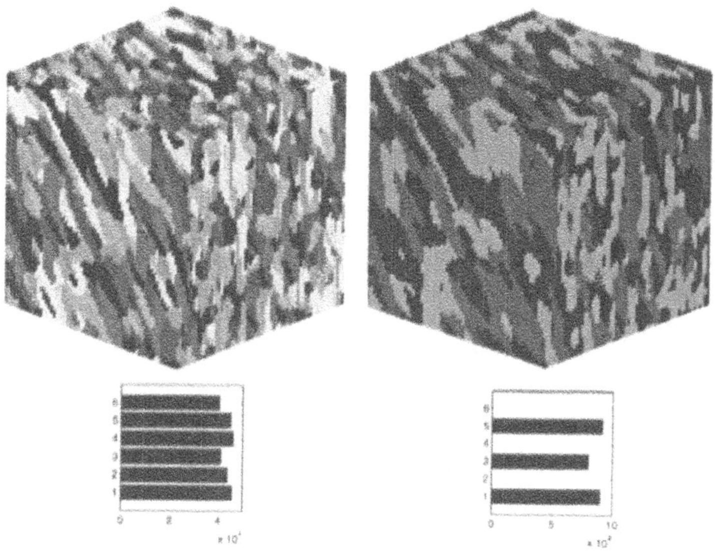

Figure 6 (color): (Right) Domain configurations of a perovskite ferroelectric after crossing the Curie point. Distinct colors refer to distinct domains. The volume fractions of each of the six domains are shown in the bar chart beneath. (Left) Domain configurations after application of an external electric field in the [111] direction. Group theoretically it is predicted domains (1,3,5) will be roughly equal[26]. This is verified in the simulation.

As an example of how to use the DAE guidance shown in Table I, Fig. 6 also shows the (a,a,a,a,a,a) to (a,b,a,b,a,b) domain shift by applying a DC poling field in the [111] direction as indicated in the third row of Table I. As seen in Fig. 6 in the bar chart below on the right, domains 1, 3, and 5 are all roughly 8000 and represent constant 'a' while domains 2, 4, and 6 are all 0 and represent constant 'b'. Many other examples of perovskite DAE have been simulated the other rows in Table I and for orthorhombic and rhombohedral child phases[21].

DISCUSSION

This work has focused on recent advances in computational ceramic modeling at the micro- and mesoscale and has discussed some of the issues surrounding each regime. The work on anion doping in TiO_2 underscores the interplay of a narrowed bandgap, overlap with oxygen orbitals, and delocalization of valence band edge states in producing efficient photocatalysts. Electronic density of states and wavefunction localization calculations for phosphorous doping in anatase indicate phosphorous is a non- to semi-efficient photocatalyst at high concentrations. Reviewing similar calculation for nitrogen, sulfur, and carbon show sulfur and low-concentration carbon and nitrogen are poor photocatalysts and high-concentration carbon and nitrogen are better photocatalysts.

The influence of defects such as vacancies in TiO_2 has also been discussed and preliminary results for the free energy of oxygen vacancy formation have been presented. Internal consistency checks show the nonequilibrium MD runs have not completely converged but are accurate enough to be nearly linear and monotonically decreasing as found previously for simple metals such as Ni[16]. Further calculations will yield increasingly accurate results and may provide insight into similar processes such as hole formation during photoexcitation reactions. This work is significant as it is the first time free energies of vacancy formation have been calculated using the power and accuracy of explicit wave functions and a DFT formalism.

Domain average engineering (DAE) of mesoscopic multidomain structures undergoing kinetically controlled ferroelectric phase transformations has been discussed. A phase field free energy functional which adequately captures the physics of such processes was presented. Group theoretical guidance for the BaTiO3-type cubic to tetragonal multidomain structure was reviewed and a simulation of DAE to one of the 12 possible[26] multidomain structures presented. Such simulations are significant because they validate abstract theoretical derivations in a empirically benchmarked simulation and are a first step in realizing the fine-grained control over such multidomain structures for enhanced optical, piezoelectric, or electrical properties, to name a few.

CONCULSION

This work has surveyed a variety of ceramic modeling situations. The techniques employed in each framework will become increasingly intertwined as true multiscaling modeling of ceramic structures emerges. As computational resources proliferate and costs decrease, such techniques will become more accurate and convenient and will find an increasingly large application to ceramics development and manufacturing.

REFERENCES

[1]J. P. Lewis, K. R. Glaesemann, G. A. Voth, J. Fritsch, A. A. Demkov, J. Ortega, O. F. Sankey, *Phys. Rev. B*, **64**, 195103, (2001).
[2]O. F. Sankey and D. J. Niklewski, *Phys. Rev. B*, **40**, 3979, (1989).
[3]G. A. Adams, M. O'Keefe, A. A. Demkov, O. F. Sankey, Y. M. Huang, *Phys. Rev. B*, **49**, 8048, (1994).
[4]A. A. Demkov, J. Ortega, O. F. Sankey, M. P. Grumbach, *Phys. Rev. B*, **52**, 1618, (1995).
[5]O. F. Sankey, A. A. Demkov, W. Windl, J. Fritsch, J. P. Lewis, M. Fuentes-Cabrera, *Int. J. Quant. Chem.*, **69**, 327, (1998).
[6]P. Jelinek, H. Wang, J. P. Lewis, O. F. Sankey, and J. Ortega, *Phys. Rev. B*, **71**, 235101, (2005).

[7]J. K. Burdett, T. Hughbanks, G. J. Miller, J. W. Richardson, and J. V. Smith, *J. Am. Chem. Soc.*, **109**, 3639, (1987).
[8]H. Wang and J. P. Lewis, Submitted, (2005).
[9]J. Pascual and H. Mathieu, *Phys. Rev. B*, **18**, 5606, (1978).
[10]R. Sanjines, H. Tang, H. Berger, F. Gozzo, G. Margaritondo, and F. Levy, *J. Appl. Phys.*, **75**, 2945, (1994).
[11]R. Asahi, T. Morikawa, T. Ohwaki, K. Aoki, and Y. Taga, *Science*, **293**, 269, (2001).
[12]J. P. Lewis, T. E. Cheatham, E. B. Starikov, H. Wang, O. F. Sankey, *J. Phys. Chem. B*, **107**, 2581, (2003).
[13]L. Lin and W. Lin and Y. Zhu and B. Zhao and Y. Xie, *Chem. Lett.*, **34**, 284, (2005).
[14]T. Umebayashi, T. Yamaki, S. Tanaka, and K. Asai, *Chem. Lett.*, **32**, 330, (2003).
[15]H. Wang and J. P. Lewis, *J. Phys.: Condens. Matter*, **17**, L209, (2005).
[16]C. R. A. Catlow and R. James, *Proc. R. Soc. London, Ser. A*, **384**, 157, (1982).
[17]M. Li, W. Hebenstreit, U. Diebold, A.M. Tyryshkin, M.K. Bowman, G.G. Dunham and M.A. Henderson, *J. Phys. Chem. B*, **104**, 4944, (2000).
[18]V. E. Heinrich and P. A. Cox, *The Surface Science of Metal Oxides*, Cambridge University Press, Cambridge, (1994).
[19]J. B. Keith, H. Wang, and J. P. Lewis, In preparation, (2005).
[20]D. Frenkel and B. Smit, *Understanding Molecular Simulation*, Academic Press, San Diego, (1996).
[21]E. Smargiassi and P. A. Madden, *Phys. Rev. B*, **51**, 117, (1995).
[22]M. de Koning, S. R. de Debiaggi, and A. M. Monti, *Phys. Rev. B*, **70**, 054105, (2004).
[23]L.Q. Chen, *Annu. Rev. Mater. Res.*, **32**, 113, (2002).
[24]W. Zhang and K. Bhattacharya, *Acta. Mat.*, **53**, 185, (2005).
[25]R. Ahluwalia and W. Cao, *J. Appl. Phys.*, **93**, 537, (2001).
[26]D. M. Hatch, H. T. Stokes, and W. Cao, *J. Appl. Phys.*, **94**, 5220-27 (2003).
[27]J. B. Keith and D. M. Hatch, In preparation, (2005).

Eco-Materials Processing and Design
Edited by Koji Watari and Soo-Wohn Lee

PREPARATION OF TiO_2 COATING ON GLASS SUBSTRATE BY NON-AQUEOUS SOL AND COMPARISON OF CONTROLED ATMOSPHERE AND MICROWAVE DRYING TECHNIQUE

Boonchoy Soontornworajit, Thammarat Panyathanmaporn, Angkhana Jaroenworaluck, and Sitthisuntorn Supothina.
National Metal and Materials Technology Center, National Science and Technology Development Agency, 114 Thailand Science Park, Paholyothin Rd., Klong Luang, Pathumthani 12120 Thailand

ABSTRACT

Interest of TiO_2 film coating on various substrates for photocatalytic application has been increased recently. In this study, non-aqueous sol was prepared using tetrapropyl orthotitanate ($Ti(OPr)_4$) and 1,2-butadiol ($Bu(OH)_2$) as starting chemicals for dip coating on glass substrate. From this synthesis method, nano-sized TiO_2 having pure anatase phase were prepared and characterized by using an x-ray diffractometer (XRD) and a transmission electron microscope (TEM). To apply the synthesis process for TiO_2 film coating, a commercial nano-sized TiO_2 powder, Degussa P25, was added and homogeneously mixed into the as-prepared sol and used as a source of TiO_2 nanoparticles. Surfaces of soda-lime glass substrate used in this study were cleaned, polished and chemically etched before coating. To control film thickness, multi-dipping of the substrate into the sol before gelatin process was applied. Before calcination process, the gel film was dried by two different techniques: drying at atmospheric condition and heating by microwave at 800 W for 2 minute. The resulting TiO_2 layers were then characterized by field emission scanning electron microscope (FE-SEM) to identify the coating characteristics and thickness, respectively. Phase present on the coating layers was revealed by XRD. From this coating technique, uniformly distributed TiO_2 islands could be prepared. Such microstructure could be suitable for gas-phase reactions.

INTRODUCTION

Titanium dioxide (TiO_2) has been used in many applications: chemical synthesis, solar energy utilization, and environmental remediation, depending on its crystal structure and surface morphology.[1-3] TiO_2 appears in two common forms, anatase and rutile, and a rare form, brookite.[2] By thermodynamic driving force, anatase transforms to rutile, the most stable form, under all conditions of temperature and pressure. Comparing those two forms, it is found that rutile structure is densely packed than those of anatase. It can be obviously confirmed by the density of rutile and anatase which is 4.23 and 3.87 respectively.[4]

The anatase-rutile transformation has been governed by many factors such as synthetic method, atmosphere, and precursor. To obtain the desirable product, many synthetic routes, with different reagents, solvents, and condition, have been conducted. The common solvents are alcohol, carboxylic acid and water. Moreover, some reagents have been added into the solvent in order to monitor the reaction. For example, inorganic acid acts as peptizer, acetic acid acts as stabilizer, and polyethylene glycol acts as gelling agent[5].

In photocatalytic applications, the photocatalytic reactions are carried on aqueous phase and gas phase reactions. TiO_2 powder dispersed in water was conventionally used for aqueous phase reactions. However, when the reaction was completed, it was difficult to separate the TiO_2

particles from the suspension. In gas phase reactions, it requires the reactors that desired gases are able to be simultaneously illuminated by light source and be contacted to the catalyst. To solve the problem of catalyst separation in aqueous reactions and to construct the reactors for gas phase reaction, TiO_2 particles were fixed on the immobilized or massive substrate such as glass silica and alumina.[6] The coating layers of TiO_2 were preceded by dipping, casting, spraying, etc. Comparing many coating method, it is realized that the dipping process is the simplest and economical process.

In this report, coating layers of TiO_2 was prepared. The effects of drying process, coating layer, and P25 content to the uniformity of coating surface were investigated.

EXPERIMENT

Chemicals

Chemical reagents used in this study include tetrapropyl orthotitanate ($Ti(OPr)_4$, Fluka), 1,2-butadiol (1,2-$Bu(OH)_2$, Fluka), tetraethyl orthosilicate (TEOS, Fluka), and hydrochloric acid (HCl, Fluka). They are all reagent grade of more than 98% purity. The P25 TiO_2 was purchased from JJ Degussa, Thailand.

Glass Substrate Preparation

Glass slides were soaked in a cleaning solution (HNO_3:H_2O_2:H_2O = 1:1:6), at room temperature for 10 minutes. The glass slides were then rinsed with DI water and acetone respectively. In order to improve the adhesion between the TiO_2 layer and the glass substrate, the silica binding layer was deposited onto the glass surface by dipping the substrate in the silica sol. The silica sol was prepared by mixing TEOS in 0.1M HCl in ethanol. The final volume was then adjusted to be 50 ml. The solutions were vigorously stirred for 2 hours and kept in a closed bottle under ambient condition for 24 hours before coating onto the glass substrate. The glass substrates were repeatedly dipped in the silica solution for three cycles and calcined at 400°C for 4 hours.

Preparation of Ti Precursors

The Ti precursor was prepared by slowly pouring a 1,2-$Bu(OH)_2$ into a $Ti(OPr)_4$. The weight ratio of the $Ti(OPr)_4$ and 1,2_BuOH was 3:1.75. The aliquot was diluted with 30 ml 2-propanol, and also contained 0 (control), 1, 2, and 5 wt% P25 compared to initial Ti source. These solutions will be referred to as Ti_0P25, Ti_1P25, Ti_2P25, and Ti_5P25, respectively.

TiO2 Coating Procedure

The glass substrates coated with a silica binding layer were dipped into the Ti precursor solutions at the depth 0.5 cm, and were withdrawn within five seconds. The TiO_2-coated glass specimens were divided into two groups. The first group was dried under ambient condition and then was re-dipped into the Ti solution. The later group was subjected to heat treatment in a microwave oven (SHARP, R-80MM) at 800 W for 2 minutes before repeating the dipping process. All the TiO_2-coated glass slides were kept in the seal containers for a week and then were calcined at 400°C for four hours.

To be able to recognize the specimen, the specimen code was used. For example, the name "mw-0P25_2L" describes the specimen that glass substrate was dipped into the Ti_0P25 solution for TiO_2 coating for two layers, and was dried in a microwave oven. Similarly, a name "am-1P25_1L" describes the glass substrate that was dipped into a Ti_1P25. for one layer, and was dried at ambient condition. The defined code was applied to all coating specimens.

Characterizations

The phase composition of the deposit layer was determined by using an X-ray diffraction technique (XRD, JDX3530), while its morphology was investigated by using a scanning electron microscope (SEM, JSM 5410).

RESULTS AND DISCUSSION

TiO2 Coating on Glass Substrate

In general observation of TiO_2 coating and atmospheric drying, it was found that TiO_2 coating showed white and opaque surface appearances. Most of the specimens showed smooth and uniform surface although cracking was visually observed in some specimens. However, when observed under electron microscope, microcracking was found as shown in Figure 1(a). The micro-cracking structure was formed by the agglomeration of nano-size TiO_2 particles, uniformly attaching onto glass substrates as shown in Figure 1(b).

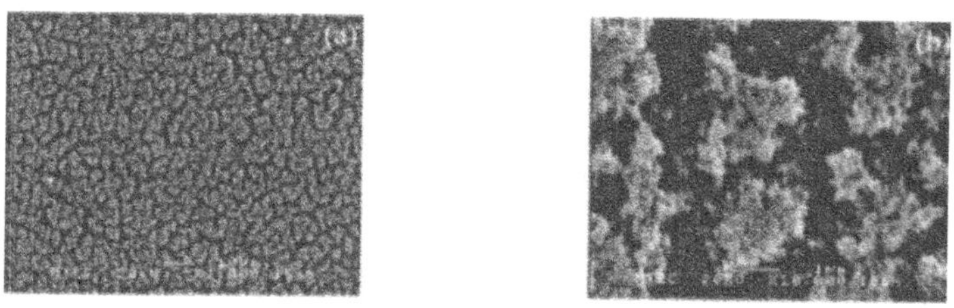

Figure 1 SEM images of the 1 layer, TiO2 coating without P25 addition. Although it is visually continuous and uniform, higher magnification (Fig. b) shows the micro-cracking structure (specimen "am-0P25_1L").

When the specimens whose SEM images are shown in Fig. 1 were recoated for 1, 3, and 5 cycles, microcracking was more visually apparent when the coated layers became thicker, particularly for 5 cycle coating. The microstructures of these specimens are shown in Figure 2. By visual observation, their appearances, however, were still smooth surface. By comparing the number of specimens which showed apparently cracked, fairly cracked, and smooth surface (Figure 3), we found that repeated coating for up to 3 cycles results to the TiO_2 layers having fairly smooth and uniform surface.

Figure 2 SEM image of the TiO2 layer deposited for (a) 1 cycle (am-0P25_1L), (b) 3 cycle (am-0P25_3L), and (c) 5 cycle (am-0P25_5L).

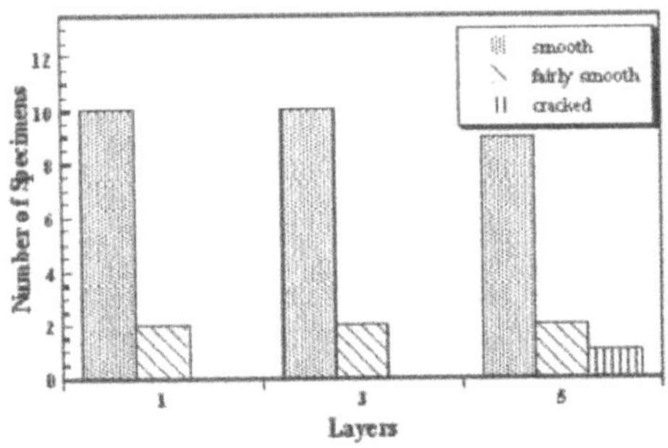

Figure 3 Correlation of coating quality (visually smooth, fairly smooth, and cracked) and the number of coating cycles (specimens "am-0P25_xL").

To improve the uniformity of the TiO_2 coating layers, a P25 powder was added into Ti sol to act as nucleation sites and reduce the shrinkage of the coating layers during drying. Figure 4 shows the SEM images of the TiO_2 coating (3 coating cycles) from the sol containing 1 – 5 wt% P25 powder. It's evident that addition of a P25 powder into the sol resulted to denser and more uniform layer. However, too much of P25 content, *i.e.* > 5wt%, caused more cracking. Figure 5 illustrates the relationship between the quality of the coating and the amount of the P25 powder added into the sol. The number of specimens that had cracked and fairly smooth surface was decreased from three to one specimen as the P25 content increased. It is able to interpret that P25 tends to reduce the crack that able to be seen by visual observation. By comparing Figure 2(b), and Figure 4(a)-(c), it is obvious that addition of P25 powder reduced the size of micro cracking plate into the micro cracking agglomeration, resulting to the uniform coating layer, especially for the thick coating.

Figure 4 SEM images of the TiO2 layer deposited from the sol containing P25 powder for (a) 1 wt% (am-1P25_3L), (b) 2 wt% (am-2P25_3L), and (c) 5 wt% (am-5P25_3L).

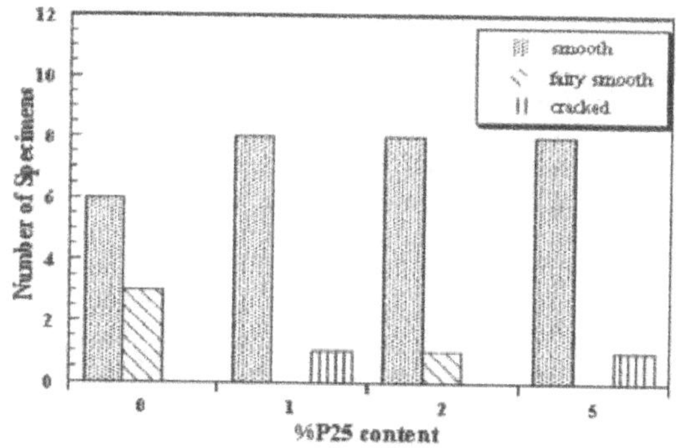

Figure 5 The number of specimens that were visually smooth, fairly smooth, and cracked. They were prepared from the sol containing 0 – 5 wt% P25 powder

Figure 6 is an XRD patterns of the TiO_2 coating layer containing various amount of P25 powders. Regardless of the amount of P25 added, the TiO2 layers had anatase structure. Note that no characteristic peaks of rutile phase of the P25 TiO_2 was observed since the quantity of the P25 added to the sol was too small.

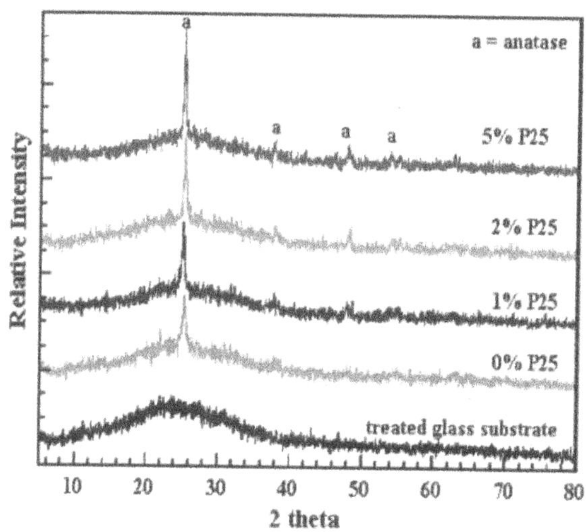

Figure 6 XRD pattern of the TiO_2 coating on glass substrate. The substrates were dipped into the sol containing various amount of P25 powder.

Comparison of Atmospheric and Microwave Drying

After dipping, the resulting specimens were subjected to drying either by leaving the specimens in the atmosphere-controlled chamber, or by means of microwave heating. By comparing the quality of the dried specimens, the number of smooth-surface specimens dried under controlled atmospheric condition (referred to as "unwaved") was twenty eight specimens, while the number of the smooth-surface specimens dried by microwave heating was found only four specimens. Moreover, the specimens dried under atmospheric condition that showed cracked surface were in less quantity than the specimens dried by microwave heating as shown in Figure 7. The microstructure of the TiO_2 coating dried at under controlled atmospheric condition was a uniform agglomeration of nano-TiO_2 particles as shown in Figures 1, 2 and 4, while that of TiO_2 dried by microwave heating showed the large plate of TiO_2 layer of approximately 20 μm in size as shown in Figure 8.

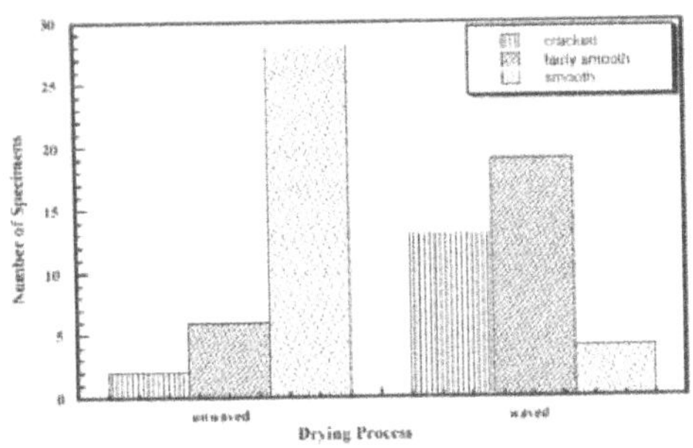

Figure 7 The number of specimens that were visually smooth, fairly smooth, and cracked after drying under controlled atmosphere (unwaved), and microwave drying (waved). They were prepared by dipping the glass substrates into the pure sol and the sol containing 1 – 5 wt% P25 powder.

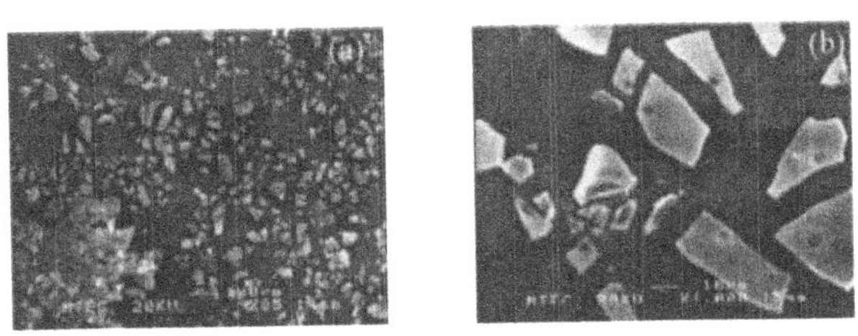

Figure 8 SEM images of the TiO2 layer prepared by dipping the glass substrates into the TiO_2 sol for 3 cycles and then subjected to microwave drying (specimen "mw-0P25_3L").

CONCLUSION

The uniform island coatings of TiO_2 were successfully prepared by controlling the number of coating cycles. The coating thickness was varied by repeating a dipping cycle which could be performed up to 5 cycles. Thicker layer resulted to large cracking due to volume shrinkage. Addition of P25 TiO_2 powder of up to 5 wt% into the sol improved the coating quality that the large cracking was reduced into microcracking. The TiO_2 layer dried under controlled atmospheric condition consisted of nano-size TiO_2 agglomeration uniformly distributed over the substrate surface. Microwave drying, on the other hand, resulted to large cracking.

ACKNOWLEDGEMENT

The author would like to thank Suphamongkol Buama, Niramol Uputchatai, and Chotchaung Polchaingdee of Burapha University, Thailand, for their technical assistants on part of this work. This research is financially supported by the National Metal and Materials Technology Center, grant# MT-B-47-CER-07-169-I.

REFERENCES

[1]Y. Zhang, A. Weidenkaff, A. Reller, " Mesoporous Structure and Phase Transition of Nano crystalline TiO_2" *Materials Letters* **54,** 375-381 (2002).

[2]K.A. Farrell, MS. Thesis in material science and engineering, Worcester Polytechnic Institue (2001).

[3]Y.V. Kolen;Ko, V.D. Maximov, A.V. Garshev, P.E. Meskin, N.N. Oleynikov, B.R. Churagulov, " Hydrothermal Synthesis of Nanocrystalline and Mesoporous Titania from Aqueous Complex Titanyl Oxalate Acid Solution" *Chemical Physics Letters* **388**, 411-415 (2004).

[4]T.C. Patton, Properties and Economies, Pigment Handbook, Jonh Wieley& Son, New York, v1 (1973).

[5]T. Trung, C.S. Ha, "One-Component Solution System to Prepare Nanometric Anatase TiO_2" *Materials Science & Engineering C* **24,**19-22 (2004).

[6]R.S. Sonawane, B.B. Kale, M.K. Dongare, "Preparation of Photo-Catalytic activity of Fe-TiO_2 Thin Films Prepared by Sol-Gel Dip Coating" *Mat. Chem. and Phys.* **85**, 52-57 (2004)

Author Index

Printed and bound by CPI Group (UK) Ltd, Croydon, CR0 4YY

16/09/2026

14963458-0001